CONSTRUCTION INDUSTRY TRAI

ESSENTIAL ELECTRICS FOR PLUMBERS, GAS FITTERS AND H & V ENGINEERS

STUDY NOTES
January 1994

© **Construction Industry Training Board 1993**   EE114

Published by:
Construction Industry Training Board
First Edition, 1994
ISBN 1 85751 090 9
© Construction Industry Training Board, 1994

CONTENTS

Module No:

1. Electrical Concepts

2. Supply Systems

3. Instrumentation

4. Regulations and Standards

5. Electrical Isolation

6. Earthing Arrangements and Protective Conductors

7. Electrical Cables and Accessories

8. Drawings and Circuit Diagrams

9. Tools and Installation Practices

10. Electrically Operated Controls and Fault Finding

11. Circuit Design

12. Inspection and Testing

ACKNOWLEDGEMENTS

The Construction Industry Training Board wishes to express its thanks to the following organisations for their co-operation in allowing extracts and illustrations from their various publications to be reproduced in these Study Notes:

Honeywell

The British Standards Institution

We also wish to thank the following companies for permission to illustrate certain of their products:

Crabtree Electrical Industries Ltd.
G.E.C. Fusegear Ltd.
Clare Instruments Ltd.

NOTE:

These Study Notes contain abbreviated extracts and paraphrases of the IEE Regulations. It is emphasised that these interpretations of the Regulations have been devised for the purpose of training and should not be regarded as authoritative in any other context. When necessary, the Regulations should be referred to directly.

Electrical Concepts

Magnetism

Most people are familiar with magnetic phenomenon. A bar of iron (or steel or some other metal) is magnetised, having a North seeking pole at the one end, and a South seeking pole at the other. Metal objects are attracted to either end, but if such objects are themselves magnetised, it will be seen that unlike poles (N & S) are attracted to each other. Equally, like poles repel.

Magnetism

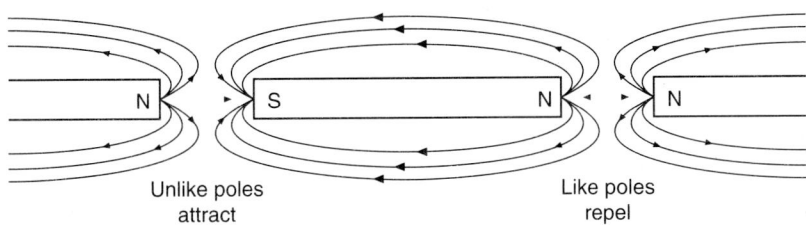

Unlike poles attract

Like poles repel

Electricity

Michael Faraday discovered how to make electricity in 1831 when he plunged a bar magnet into a coil wire and generated a wave of electricity. He later found that by rotating a copper plate between the poles of a magnet, power could be taken from the axis to the rim of the disk.

The system of holding the coil of wire stationary while varying the strength of the magnetic field is used in power stations. The bar magnet is replaced by a rotating electro-magnet and the coils are arranged so that the windings are cut by the magnetic field as the magnet rotates.

In a power station the magnets in each generator are turned by steam-driven turbines which change mechanical energy into electrical energy.
The rate of exchange is 1 horse power 746 watts therefore a 1KW electric fire requires the equivalent of 1 1/3 horsepower.

Electrical Pressure (Voltage)

Any equipment that works by electricity needs a power supply. Different equipment needs different types of supply. A torch may require two 1.5 volt batteries, whilst an electric shower needs a 240 volt supply.

Force

Voltage is the force behind electricity; it is often referred to as electric pressure and can be readily compared with the water pressure in a plumbing system. The pressure which drives the water is due to the difference of levels between the tank and the tap. The difference in the voltage levels between two points is called the potential difference (p.d.)

The unit of measurement for electrical pressure (electro-motive force or emf) is the volt. The symbol used is V.

Very high voltages which are used on power lines carried on pylons are measured in kilovolts, symbol kV, where one kilovolt is equal to 1,000 volts.

Types of Power Supply

There are two types of power supply, alternating current known as a.c. and direct current known as d.c. Alternating current is the type of electricity supplied to domestic, commercial and industrial premises by the local electricity companies. Direct current is the type of electricity you get from a battery or by using certain components to form a special type of circuit. It is possible to change a.c. to d.c. and vice-versa.

Electric Current (Amperes)

In a conventional circuit, an electric current, unlike water, will only flow if it can return to its source. The route it takes is known as a circuit. If you break a circuit by cutting a wire forming that circuit, the current stops.
The unit of current is the ampere and is measured in amps (A). An immersion heater takes around 12A and a shower about 24A, whilst the electronic components in a boiler may take only 1 milliamp (mA) which is equivalent to one thousandth of an amp.

N.B. Voltage appears across components and current flows through them.

The Structure of Matter

Electrons

All materials consist of tiny particles called atoms. Materials which consist of the same kind of atom are called elements. The simplest atom is that of hydrogen. If we could see it, it would resemble the Earth with the Moon orbiting around it.

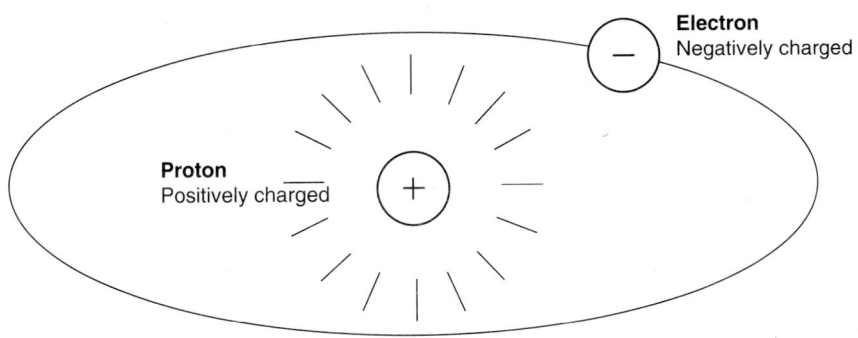

The atom consists of a central core called the nucleus and a spinning electron moving around the nucleus. The nucleus in this case consists of one positive charge of electricity called a proton, whilst the electron has an equal but negative charge.

It must be remembered that like charges repel each other and unlike charges attract each other. This rule applies to the charges of the electron and proton.

Since the nucleus is positive and the electron is negative, the electron is bound in orbit and is normally prevented from flying off because of the attraction between the two particles. The number of orbiting electrons in a given atom depends on the type of element.

Free Electrons

Consider a copper atom illustrated below. It has 29 orbiting electrons arranged in four shells.

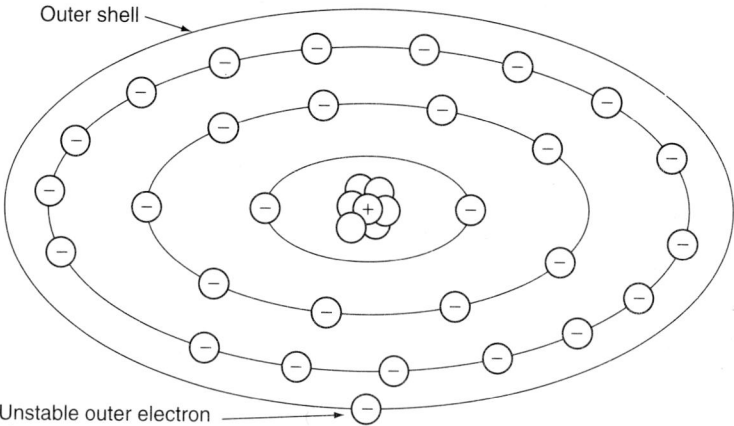

The outer electron, farthest from the nucleus, is only weakly attracted to the positively charged nucleus. This means that it can easily fly off or be dislodged. Electrons dislodged from their orbit can wander at random from atom to atom within the material and so are called free electrons. These free electrons can form the basis of an electric current.

Conductors and Insulators

Materials in which the outer electrons are not tightly bound in the atoms and can easily be dislodged to produce free electrons are called conductors, e.g. copper and aluminium. Conversely, in some materials, the orbiting electrons are so tightly bound that they cannot easily be encouraged to break away from their orbits. These materials are called insulators; insulators have virtually no free electrons available to form an electric current. Examples of insulators are plastics and ceramics.

The Random Movement of Electrons

Under normal conditions, free electrons move randomly in a conductor; the effects of temperature cause this movement (see illustration).

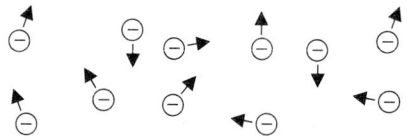

Random electron movement in a conductor

The movement is usually equal in all directions so that no electrons are given up by the material, nor are any added. This is not electric current. If however the free electrons can be encouraged to move in the same general direction within the conductor and be made to enter and leave it, this flow does constitute an electric current.

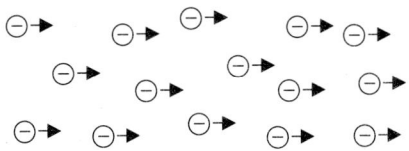

Directional movement - electric current

The principle that like charges repel and unlike charges attract can be used to force electrons to move in the same general direction and produce current flow.

This is achieved by an external negative charge (which is no more than a point with an abundance of electrons) placed at one end of a conductor and an external positive charge (a point with a deficit of electrons) placed at the other. This causes free electrons to flow towards the positive end, since electrons are negatively charged and unlike charges attract. Further, the external negative charge repels free electrons into the material so that there is ordered, directional electron flow from negative to positive as illustrated.

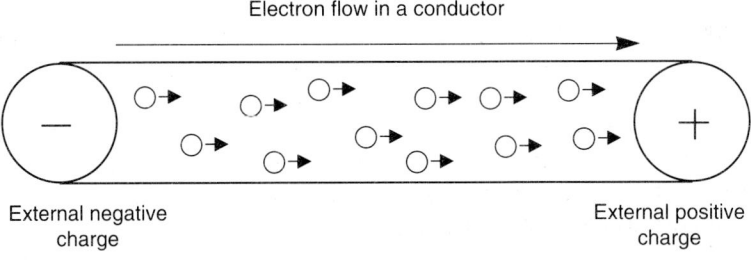

Electron flow in a conductor

External negative charge

External positive charge

In conductive materials, the electrons not tightly linked to their atoms can be dislodged from their orbit by the force from an electric power supply. In this situation some of the negatively charged electrons are free to move and flow towards the positive terminal of the power supply.

Summary

- An electric current is a flow of electrons
- A conductor is a material in which electrons are able to move freely.
- An insulator is a material in which electrons are tightly bound to their respective atoms.

Types of Conductors and Insulators

The most common types of conductors found in electrical installations are:

    Copper: Found in cable and flex
    Brass: Found in electrical accessories as terminal blocks
    Nichrome: Found in electric fire elements

The most common type of insulator used in electrical installations is plastic, of which polyvinyl chloride (pvc), a thermoplastic is the most widely used. Its wide use is due to its ability to be plasticised, giving a range of flexible plastics, from rigid to pliable. The softer material is used as insulating covering for electric cables and wiring.

Resistance

If electric current is like a flow of water, the path it flows along, which is the electrical circuit, can be likened to a heating system with obstacles in the path, like radiators, that reduce the flow of water in the system.

In electrical circuits even the circuit conductors provide some degree of resistance to the flow of current. This is why voltage is always needed to push the current around the circuit to overcome the resistance.

Consider any piece of equipment - for example, a 110 volt electric shaver. If it is accidentally connected to a 240 volt supply system too much current will flow through it and it may burn out. The solution would be to fit some kind of resistance in the circuit to limit the current. This is what happens when electric shavers are made to work on either 240/110 volt systems.

When an electric current meets resistance in a circuit it generates heat by working its way through the conductor. This heating effect can be put to a practical use when resistance elements are used in electric fires and also in the filament of a lamp where the heat produced is enough to make the filament white hot.

Resistance can cause problems in a circuit, since overheating is a major cause of electrical breakdowns and can give rise to fire risks. If, when terminating conductors into electrical accessories, or flexible cords into plug-tops, the terminal screw is not tightened sufficiently, the resulting high resistance connection will get hot and damage the accessory. Another factor to consider is the length of cable; the greater the length of run, the greater the resistance, and hence the greater the heat created. Care needs to be taken to ensure that any heat created in cables due to applied load and conductor resistance does not damage the insulation.

The resistance of a component or part of a circuit is measured in ohms, and the Omega symbol (Ω) is used to represent it.

Typical values of resistance used in electrical installation and maintenance work are:

OHM's   1k = 1,000Ω or 1 kilohm
        1M = 1,000,000 or 1 Megohm

Due to the different materials and the properties of those materials used in cables and transformers which supply a.c. electricity to a building, these are not classed as pure resistive components and their resistive element is referred to as 'impedance'. For installations up to 100 amps concerned with ring and radial circuits of power and lighting, impedance can be considered as resistance.

Relationships (Ohms Law)

If the resistance of a circuit is high, a high voltage is required to push the current round the circuit. When the voltage falls and the resistance of the circuit remains the same, there is less current. From this it is evident that the voltage, current and resistance in a circuit are related to each other. This relationship is known as Ohms Law, after Georg Ohm, a German physicist.

Expressed mathematically:

$$I = \frac{V}{R} \quad \text{current, in amps} = \frac{\text{emf (in volts)}}{\text{resistance } (\Omega)}$$

$$\text{alternatively } V = I \times R \quad \text{or } R = \frac{V}{I}$$

In terms of a simple circuit of battery and flashlamp bulb; if the battery has an emf of 12 volts and the bulb a resistance of 4 ohms, a current of 3 amperes will flow when the switch is closed.

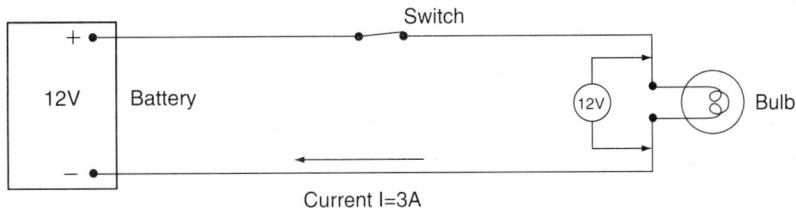

$$I = \frac{V}{R} = \frac{12}{4} = 3A$$

Series Circuits

In the previous example we were concerned with a simple direct current circuit containing only a battery and lamp. Some circuits contain a number of items, connected in series; that is to say that the same current passes through each item, in sequence.

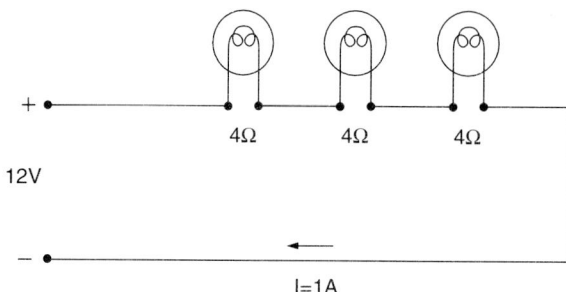

In the above circuit the same current passes through each lamp in turn. If each lamp has a resistance of 4 ohms, the total resistance of the circuit would be 12 ohms. With an emf of 12 volts, it is obvious (applying ohms law) that a current of 1 ampere would flow, and that this 1 amp current would be common to all three lamps.

The total emf of 12 volts would be distributed across the total load of all three lamps. If all of the lamps were the same, (with the same current and power rating factors) each lamp would, in effect, have 4 volts across its terminals.

This can be proved by applying ohms law, transposed as follows:

Voltage Across Lamp $V = I \times R$

$4V = 1A \times 4\Omega$

If the lamps did not all have the same resistance, different voltages would be developed across the terminals of each lamp, for example:

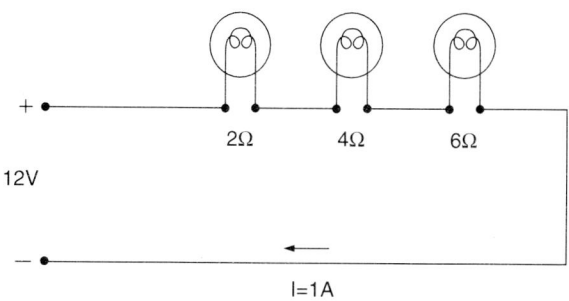

The total resistance of the circuit is still 12 ohms, and a common current of 1 ampere will flow through each of the lamps in turn. However (from ohms law) we can see that the voltage across each lamp would be different e.g.

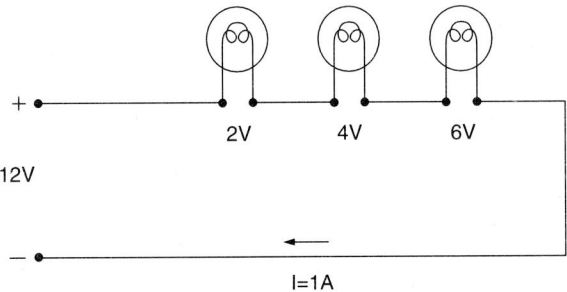

Potential Difference (p.d.)

This voltage across each of the lamps is the potential difference p.d. required to sustain the common current, (in this case) of 1 ampere.

If a similar circuit is devised, but the lamps are replaced with resistors in series, different voltages can be taken at different points, e.g.

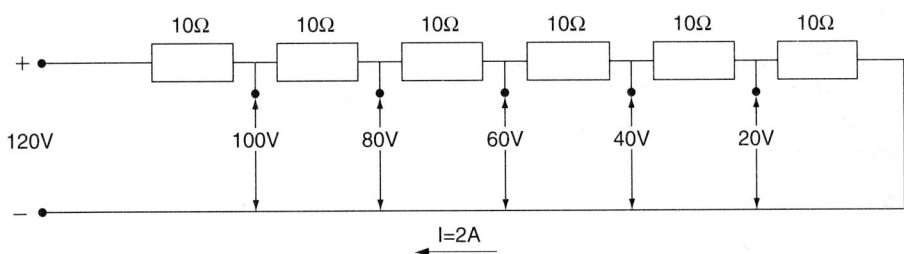

The total resistance of this circuit is 6 x 10 ohms = 60 ohms. If the input voltage is 120 volts a current of 2 amps will flow through the circuit. The total voltage of 120 volts will be "divided" across all six resistors. Since they all have the same value (10 ohms) the "voltage drop" or "potential difference" across each would be 20 volts and the voltage, measured *with respect to the common negative* would vary in 20 volt steps.

This type of circuit is called a "potential divider" network and is commonly used where different or varying voltages are required.

Variable Resistance

Where a varying voltage is required this can be achieved by placing a variable resistor in the circuit, e.g.

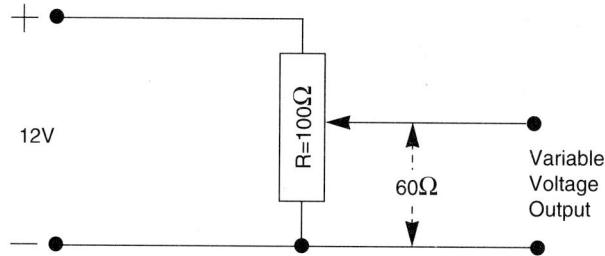

In this type of circuit, the output voltage (with no load connected) is "tapped-off" according to the position of the wiper arm. With an input current of 0.12A flowing, 7.2V will be dapped across the upper portion of the resistor, leaving 4.8V across the output terminals. This type of circuit could be used to provide a reference voltage for, say, a boiler temperature control circuit.

Parallel Circuits

Elements in an electrical circuit may also be connected "in parallel" - that is the same emf is applied to each element, but the current flowing in each will vary, depending on its resistance. The greater the resistance, the less the current in each element.

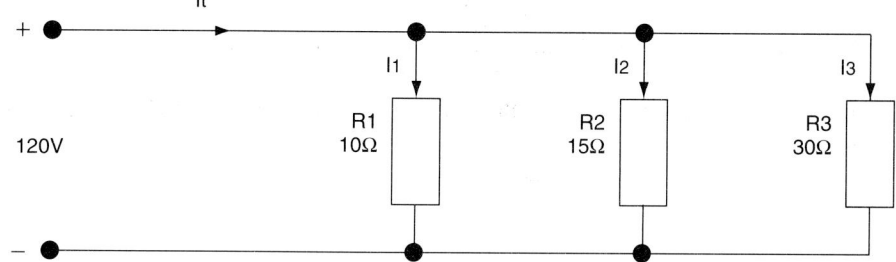

The effective resistance of such a circuit is given by the formula:

$$\frac{1}{R_t} = \frac{1}{R_1} + \frac{1}{R_2} + \frac{1}{R_3} \text{ etc}$$

$$\frac{1}{R_t} = \frac{1}{10} + \frac{1}{15} + \frac{1}{30}$$

$$\frac{1}{R_t} = \frac{6 + 4 + 2}{60}$$

$$\frac{1}{R_t} = \frac{1}{5} \qquad \therefore \frac{R_t}{1} = \frac{5}{1} = 5\Omega$$

Note: R_t = Total Resistance.

The current flowing through each element would be:

$$I_1 = \frac{V}{R_1} = \frac{120}{10} = 12A \qquad I_2 = \frac{V}{R_2} = \frac{120}{15} = 8A \qquad I_3 = \frac{V}{R_3} = \frac{120}{30} = 4A$$

If these are added together the total current (It) flowing through the circuit would be:

It   = $I_1 + I_2 + I_3$

It   =   12A + 8A + 4A

It   =   **24A**

which confirms the solution above.

    Note It = Total current

In practice, many devices are connected in parallel and it is important to know the extent of the current flowing in each.

Combined Series and Parallel Circuits

Many circuits include combinations of series and parallel circuits; an example is shown below:

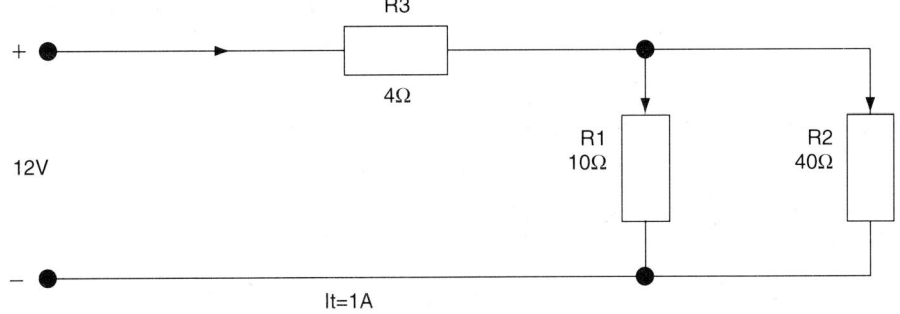

The effective resistance, and actual current flow can be determined by employing the principles and formulae previously discussed.

The resistance of the parallel element is determined by the formulae:

$$\frac{1}{R_t} = \frac{1}{R_1} + \frac{1}{R_2}$$

$$= \frac{1}{10} + \frac{1}{40}$$

$$= \frac{4+1}{40}$$

$$= \frac{5}{40}$$

$$\therefore \frac{R_t}{1} = \frac{40}{5} = 8\Omega \qquad \therefore R_t = \underline{8\Omega}$$

The parallel element is effectively in series with R_3 - so that the effective resistance of the entire circuit is:

$$4 + 8 = \underline{12\Omega}$$

From ohms law, we can determine the current flowing in the circuit would be 1 ampere. The voltage drop across the series resistance is 4 volts, with 8 volts across the parallel element.

We can now determine the current flowing in each part of the parallel element, again by the application of ohms law.

$$I_1 = \frac{8v}{10\Omega} \qquad\qquad I_1 = \frac{8v}{40\Omega}$$

$$I_1 = \underline{0.8A} \qquad\qquad I_2 = \underline{0.2A}$$

The total current passing through the parallel element is 0.8 + 0.2 = 1A confirming the result previously determined.

Electric Power

In order to do its work electricity generates power. Power is the rate at which electrical energy is being converted into other kinds of energy such as heat, light or movement in the case of electric motors.

The unit of power is the Watt and typical values of power used in electrical circuits are:

    Kilowatt = 1,000 watts 1 kW

    Megawatt = 1,000,000 watts 1MW

Many electric motors are rated in horse power 1h.p. = 746 watts. Electrical power can be calculated by multiplying the volts by the amps.

Watts(P) = volts(V) x amps(I)

It must be remembered that the formula applies accurately only to d.c. supplies, but can be used for rough calculations for a.c. circuits.

Power = voltage x current

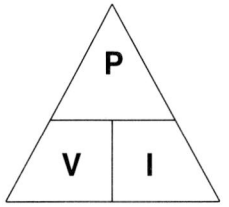

Alternatively $I = \dfrac{P}{V}$ or $V = \dfrac{P}{I}$

Example: What is the actual current taken by a 3kW immersion heater, if the supply system is 240V?

$$I = \dfrac{P}{V} = \dfrac{V = 3 \times 1000}{240} = 12.5A$$

Inductors

Electromagnetic Induction

Electrical energy is produced in a conductor by the magneto-electric effect created when a conductor is moved in a magnetic field. The creation of a voltage in a conductor by this means is termed 'induction'. We say that a current is induced in a conductor, provided that the conductor forms part of an overall circuit.

It follows that a similar result can be achieved if the conductor is stationary and the magnetic field is moved, i.e. a current is induced in the conductor as a result of the magnetic lines of flux cutting the conductor.

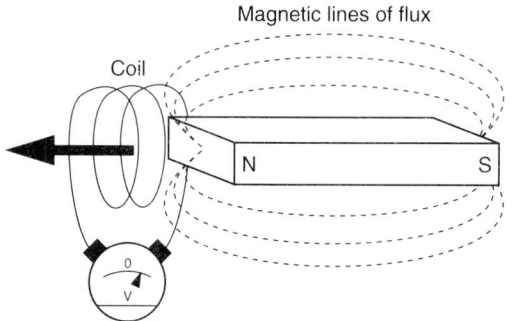

Mutual Induction

In the illustration a permanent magnet provides the magnetic lines of flux to cut the conductor. The permanent magnet could be replaced by an electromagnet. In this case, a current is passed through a coil to produce a magnetic field and the necessary flux to induce a voltage across the conductor. When a current in one coil induces a current in another adjacent coil, this is termed mutual induction.

Transformers

Transformers are inductive devices which use the properties of mutual induction. This means that they rely on a continually changing primary flux to allow an induced voltage to be developed across the secondary. Therefore they can only be used to transform a.c. voltages; they cannot work at all in pure d.c. circuits where the primary current is kept constant.

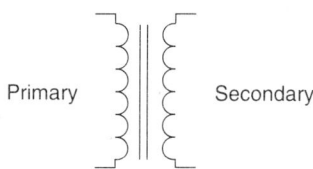

Transformers are used as voltage conversion devices, changing the level of one a.c. voltage to another either upwards or downwards. Alternatively, they can be used as isolation devices allowing two circuits to be coupled without there being a direct electrical connection as in the case of bathroom shaver units.

If the number of turns on the primary winding equals the number of turns on the secondary, then the induced secondary voltage will equal the applied primary voltage. The transformer is said then to have a 1:1 turns ratio. In practice, losses within the transformer mean that the turns ratio only gives an approximate guide to the primary : secondary voltage relationship.

A turns ratio of 10:1 would produce a voltage across the secondary coil of one tenth of the input voltage, i.e. a 240V input would produce a 24V output. Transformers can also be used to "step up" a voltage. A turns ratio of 1:4 would produce almost 1000V output from normal mains input.

Secondary windings often have a number of graduated "taps" which provide a variety of outputs. Some transformers include a centre tap in the secondary coil, which is usually earthed. This effectively halves the maximum output voltage relative to earth. In most cases the soft iron core and outer casing is bonded to earth. This type of transformer is used to supply 110V portable tools on construction sites.

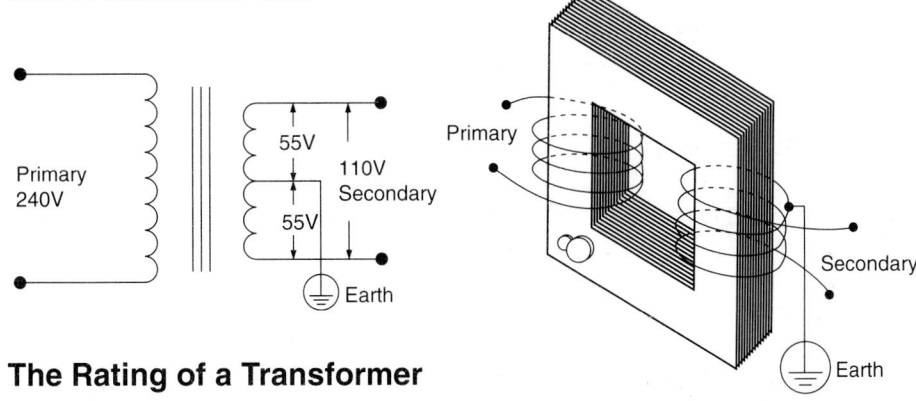

The Rating of a Transformer

With electrical equipment we must always consider the maximum current that can be carried without exceeding the rating. Transformers are limited in the amount of current they can supply from their secondary windings. If too much current is drawn, the windings get hot, possibly melting the insulation of the winding and burning it out.

Transformers are power-rated in volt-amps (VA). Watts cannot be used to represent power dissipation in a transformer because the voltage and current are not in phase with each other as they are in a pure resistance. By using a volt-amps figure which is essentially the same arithmetically speaking, it is relatively simple to calculate the current which can be drawn from a transformer.

When we say a transformer has a particular VA rating we are saying that this is the maximum power that can be drawn from the secondary/secondaries. Consider a mains step-down transformer having a VA rating of 1kVA and a single secondary winding of 110V; what maximum current can safely be drawn from the secondary?

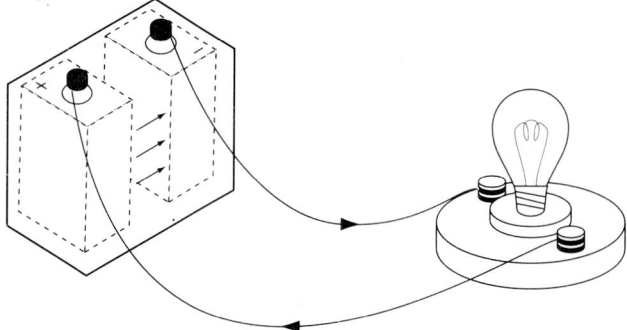

$$\text{Maximum secondary current} = \frac{\text{VA rating}}{\text{Secondary voltage}} = \frac{1000}{110}$$

$$= \underline{9\text{ A}}$$

Waveforms

The electricity generated and supplied to consumers is in the form of sinusoidal voltages which produce an alternating current. Most electronic circuits are powered from a direct current supply, the mains supply being converted into d.c. voltage.

The illustration shows a constant voltage applied across the fixed resistance of a lamp filament. The resultant current is made up of a fixed amount of electrons flowing at a constant rate from the negative to the positive terminal of the battery.

Circuit voltages can be represented pictorially. The diagram below shows the battery voltage; the vertical axis represents voltage values and the horizontal axis represents time. Since the positive terminal of the battery is perpetually 12V higher than the negative terminal, a horizontal line is drawn parallel to the time axis, indicating the voltage to be a constant value.

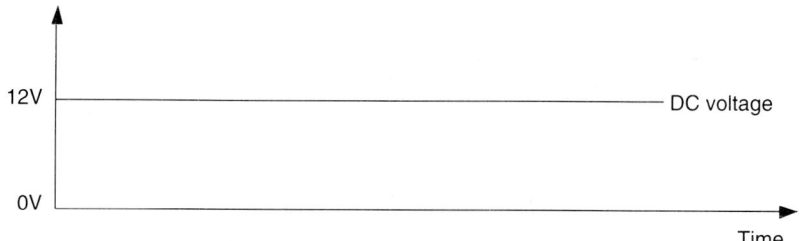

Alternating Current (a.c.)

An alternating current is made up of a flow of electrons around a circuit which continually reverses direction. This implies that there must be instances in time when the electron flow reduces to zero. The direction of the current flow is determined by the polarity of the applied voltage.

The diagram shows the mains waveform known as a sine wave. The shape of the waveform is such that the applied voltage is perpetually changing in value with respect to time, and at points 1 and 2 the supply voltage reverses polarity.

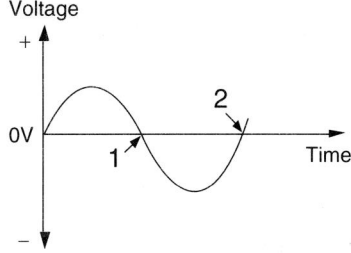

The following illustrations and diagrams show how sine waves are generated.

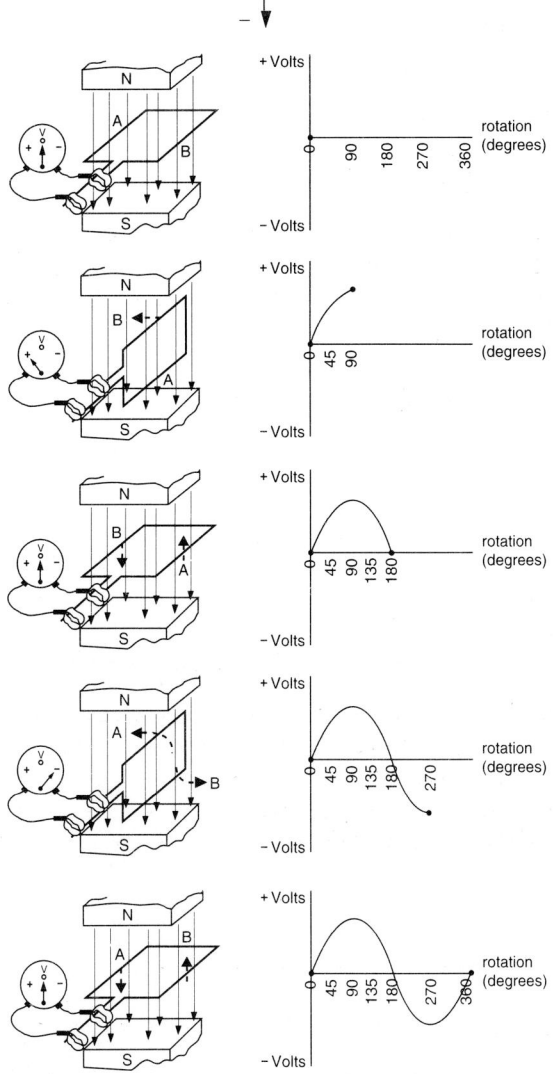

Sine Wave Values

The value of a constant d.c. voltage can only be expressed in one way. For example, when you buy a 9 volt battery it is understood that the voltage difference between the two terminals is 9 volts. The 'value' of a sine wave, however, may be expressed in several different ways because the voltage level of the waveform is continually changing.

Each sine wave is made up of an infinite number of instantaneous values. The maximum value of a sine wave is known as the peak value. Two peak values are produced during the generation of every cycle as illustrated.

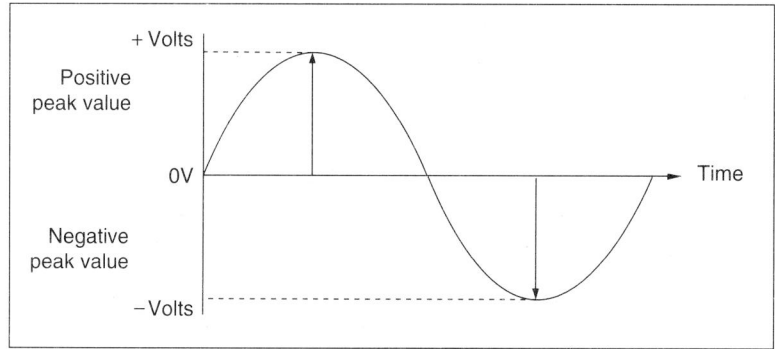

Using the Mains waveform as an example, the peak value occurs at the instant that the voltage difference between the phase and neutral conductors is at a maximum. The positive peak value is achieved when the phase conductor is at the maximum positive voltage with respect to the neutral conductor. Conversely, the negative peak value is reached when the voltage between the phase and neutral conductors is at a maximum negative value.
The peak-to-peak value is indicated on the diagram. It is the difference between the peak positive and peak negative voltages.

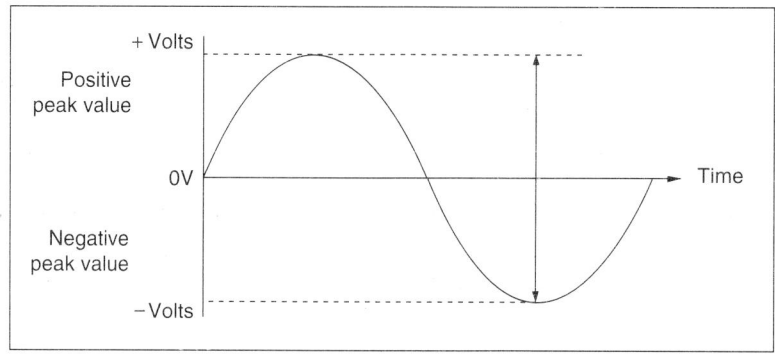

Root Mean Square (r.m.s.) Value

The root mean square value is the value most commonly used when expressing mains voltage as a value of 240 volts.

In order to explain what this means we can use, as an example, a 100W bulb which glows at a fixed brightness when connected to the mains supply.

The direction of current flowing through the filament plays no part in determining the brightness. This is determined by the power developed within the filament. The current flowing through the filament and hence the power level are continually changing as the supply voltage changes. When the mains waveform reaches a peak value, maximum current and hence maximum power are developed in the load. Similarly as the sine wave voltage reaches zero, no current and therefore no power is developed in the bulb.

If this is the case why do we not see the brightness vary? The answer is that it would if the frequency of the mains was much lower than 50Hz. Effectively the power and heat are averaged out during each cycle.

To produce the same average power in the load, the peak value of a sine wave must be higher than the r.m.s. value. The relationship between the peak and r.m.s. values has long been established and is illustrated.

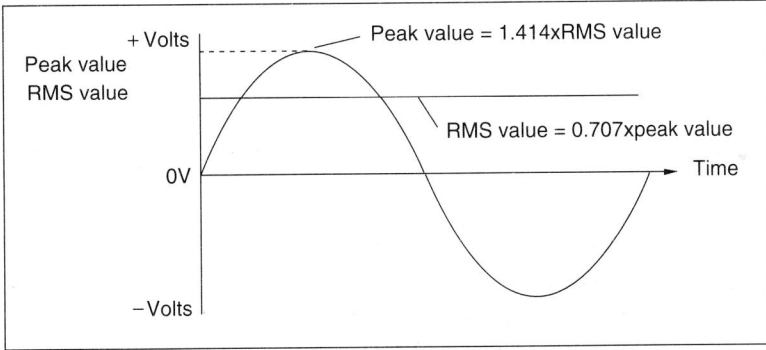

The relationship can be proved by a mathematical process. However it is beyond the scope of this publication.

Capacitance

A capacitor (sometimes referred to as a condenser) in its simplest sense is a device for the temporary storage of electrical energy. It comprises two parallel metal plates, insulated from each other. If a d.c. voltage is connected across them, one of the plates becomes rich in electrons; the other plate becomes correspondingly poor. In acquiring this 'charge' a current flows, but only for an instant. No sustained direct current can flow between the plates, since they are insulated, one from the other.

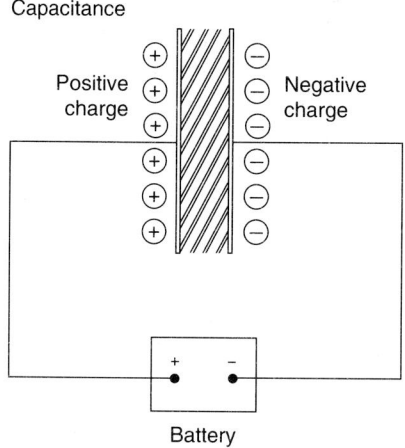

If the d.c. source is removed, the capacitor will retain its charge until it is discharged through an external circuit.

If an alternating current is fed to a capacitor it will commence to charge on one half-cycle, but as the voltage falls from its peak, will attempt to discharge, endeavouring to charge up again (in the opposite direction) on the next half cycle and so on. As a result, a capacitor appears to pass current when connected to an a.c. source, but prevents the passage of d.c. current.

The larger the area of the plates in a capacitor the greater the capacitance. In practice, a capacitor is made from two thin sheets of metal foil, insulated by waxed paper, mica, or similar material known as a dielectric.

The unit of capacitance is the farad. In practice this is far too big, and the micro-fared (one millionth of a farad) is the unit in common use. This is sometimes written as µF.

Rectifiers

It is sometimes necessary to convert an alternating current into a direct current. This is done by a process known as rectification. A rectifier is rather like a valve which permits current to pass freely in one direction but which prevents current passing in the opposite direction.

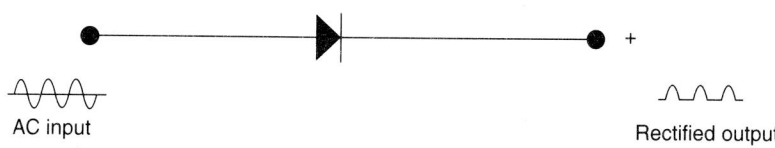

As we have seen an alternating current is one which passes through a cycle from a positive peak through zero to a peak in the opposite (negative) direction. A simple rectifier would prevent current passing on the negative cycle but would permit the positive element to pass through. The resultant output would be a series of peak voltages. This is then usually passed through some kind of filter (normally made up of an inductance coil and a capacitor) which has the effect of smoothing out the peaks, producing a steady d.c. voltage.

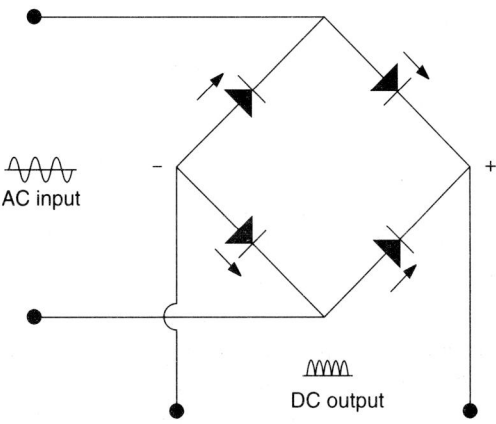

The rectifier may be a single, solid-state diode or a more complex full wave or bridge rectifier which is so arranged as to rectify both the positive and negative half-cycles of an alternating current, producing a d.c. output at twice the original frequency, which is easier to smooth (and twice the average voltage).

Rectifier devices such as this are commonly found in control equipment.

Filter Circuits

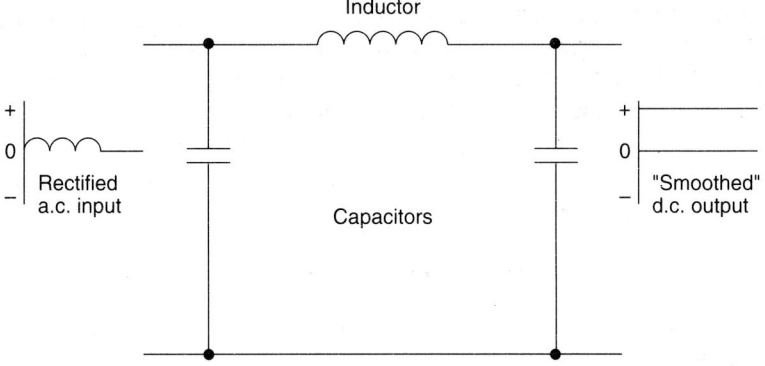

The output from a full-wave rectifier produces pulses of unidirectional current rather than a smooth d.c. output like that produced by a battery. If smooth d.c. output is required, there are various types of filter circuit which can be added, from a simple capacitor to the more common pi filter circuit shown above.

The capacitors and inductors are both devices which will store energy when it is available from the supply and release it when the supply reduces, thus producing a smoothing effect. The two types of device complement each other because the capacitor works best at smaller values of load current and the inductor works best when large values of current are being supplied. This circuit will therefore provide a smooth output across a wide range of load currents.

Three-Phase Induction Motor

When a three-phase supply is applied to windings equally distributed around the stationary part of an electrical machine, an alternating current circulates in the coils and establishes a magnetic flux which rotates at the same speed as the supply frequency. If the supply frequency were 50Hz the flux would rotate at 50Hz x 60 secs/min = 3000 rpm. This is called the *synchronous speed*.

An induction motor consists of a stator or stationary winding in which the rotating magnetic flux is established and a rotor, the rotating part of the motor. The most common type of induction motor rotor is the cage rotor, which consists of solid conductors placed in a laminated core with both ends shorted out with an end ring as illustrated.

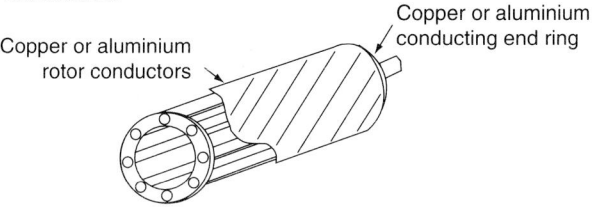

The rotating magnetic flux established in the stator windings cuts the rotor conductors and induces an emf. Because the rotor conductors are shorted, currents circulate in the rotor and set up magnetic fluxes which interact with the stator flux, and create a turning force as illustrated.

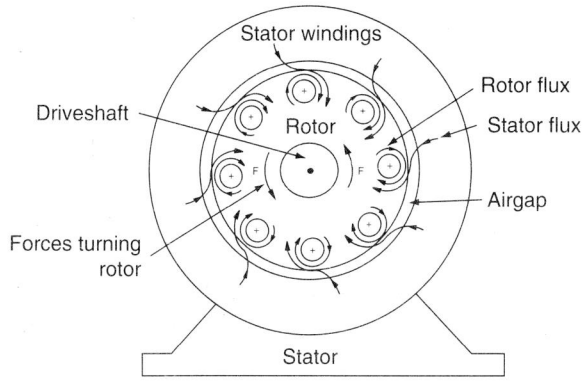

The cage induction motor has a small starting torque and should be used with light loads or started with the load disconnected. The speed is almost constant at about 5% less than synchronous speed. Its applications are for constant speed machines such as fans and pumps. Reversal of rotation is achieved by reversing any two of the stator winding connections.

Single-Phase a.c. Motors

A single-phase a.c. supply produces a pulsating magnetic field, not the rotating magnetic field produced by a three-phase supply. All a.c. motors require a rotating field to start. Therefore, single-phase a.c. motors have two windings which are electrically separated by about 90°. The two windings are known as the start and run windings. The magnetic fields produced by currents flowing through these out-of-phase windings create the rotating field and turning force required to start the motor. Once rotation is established, the pulsating field in the run winding is sufficient to maintain rotation and the start winding is disconnected by a centrifugal switch which operates when the motor has reached about 80% of the full load speed.

A cage rotor is used on single-phase a.c. motors, the turning force being produced in the way described previously for three phase induction motors. Because both windings carry currents which are out of phase with each other, the motor is known as a 'split-phase' motor. The phase displacement between the currents in the windings is achieved in one of two ways:

* by connecting a capacitor in series with the start winding, as illustrated, which gives a 90° phase difference between the currents in the start and run windings.
* by designing the start winding to have a high resistance and the run winding a high inductance, once again creating a 90° phase shift between the currents in each winding as illustrated.

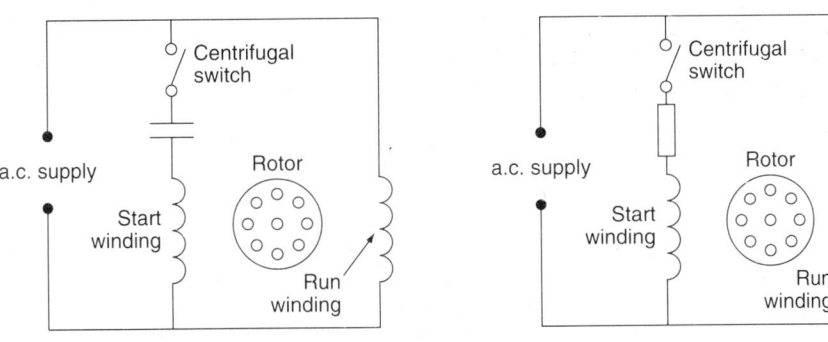

(a) Capacitor split-phase motor   (b) Resistance split-phase motor

When the motor is first switched on, the centrifugal switch is closed and the magnetic fields from the two coils produce the turning force required to run the rotor up to full speed. When the motor reaches about 80% of full speed, the centrifugal switch opens and the machines continues to run on the magnetic flux created by the run winding only. Split-phase motors are constant speed machines with a low starting torque and are used on light loads such as fans, pumps, refrigerators and washing machines. Reversal of rotation may be achieved by reversing the connections to the start of run windings, but not both.

Shaded Pole Motors

The shaded pole motor is a simple, robust, single-phase motor, which is suitable for very small machines with a rating of less than about 50 watts. Illustrated below is a shaded pole motor. It has a cage rotor and the moving field is produced by enclosing one side of each stator pole in a solid copper or brass ring, called a shading ring, which displaces the magnetic field and creates an artificial phase shift.

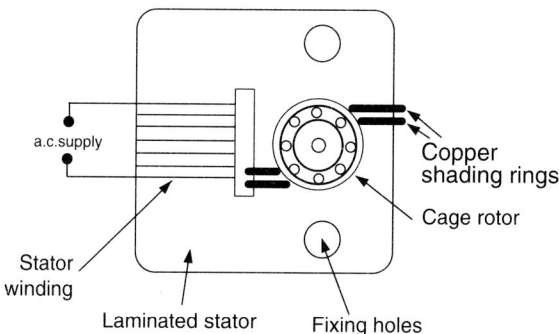

Shaded pole motors are constant speed machines with a very low starting torque and are used on very light loads such as oven fans and electric fan heaters. Reversal of rotation is theoretically possible by moving the shading rings to the opposite side of the stator pole face. In practice this is often not a simple process and since the motors are symmetrical it is sometimes easier to reverse the rotor by removing and fixing bolts and reversing the whole motor.

There are more motors operating from single-phase supplies than all other types of motor added together. Most are used in very small motors in domestic and business machines, where single-phase supplies are common.

Motor Maintenance

All rotating machines are subject to wear, simply because they rotate. Motor fans which provide cooling also pull dust particles from the surrounding air into the motor enclosure.

Bearings dry out, drive belts stretch and lubricating oils and greases require replacement at regular intervals. Industrial electric motors are often operated in a hot, dirty, dusty or corrosive environment for many years. If they are to give good and reliable service they must be suitable for the task and the conditions in which they operate. Maintenance is required at regular intervals.

The solid construction of the cage rotor used in many a.c. machines makes them almost indestructible, and, since there are no external connections to the rotor, the need for slip rings and brushes is eliminated. These characteristics give cage rotor a.c. machines maximum reliability with the minimum of maintenance and make the induction motor the most widely used in industry. Often the only maintenance required with an a.c. machine is lubrication in accordance with the manufacturer's recommendations.

Supply Systems

History of Development

In 1878, Thomas Edison developed the first electric light bulb. It was a marvellous invention, but Edison had to be able to get electricity to the users, so that they could buy and use his new light. At this time Edison was using direct current (d.c.) electricity, which is transmitted by two wires, one 'positive' and one 'negative'. Unfortunately, he found that the wires he used to carry the electricity had a resistance to the current, and if the bulb was more than about 3km away from the power station, the light it produced was too dim to be of any use. In order to provide everyone with a sufficiently bright light, Edison would have had to build power stations every six kilometres.

In 1885, Westinghouse and Stanley's experiments showed they could reduce the energy loss by using alternating current and a transformer. It took Edison twenty years to concede that higher efficiencies could be achieved by a.c. power transmission.

The basic principles used by Westinghouse and Stanley are still used today in our National Grid. The main problem is that wires have resistance. When a current passes through a resistor, heat is generated and some of the energy from the electricity is lost as heat. We can reduce this heat loss in one of two ways. The resistance of the wire can be reduced by making it thicker, but this would cost more and use up valuable resources of copper (or other metals). The other solution is to reduce the amount of current flowing.

To work out the power transferred by an electric current we used the formula:

Power transferred = Voltage supplied x Current flowing.

Clearly, we want to keep the power transferred the same, so in order to reduce the current we must increase the voltage. With d.c. electricity this is difficult, but with a.c. we can use a transformer.

By having more turns in the output (secondary coil) of the transformer, we can increase the voltage and therefore reduce the current. When we need to reduce the voltage, we pass the electricity through a transformer with fewer wires in the secondary coil than in the primary coil. This reduces the voltage and increases the current.

The Distribution System

The power generating companies use step-up transformers to increase the voltage on the overhead cables to 275,000V (275kV) or, in some cases, 400,000V (400kV). The electricity is transported across country to towns and cities, often by overhead cables connected to pylons. The voltage is reduced to the required level by local transformers, which are usually called substations.

Electricity is usually supplied to the home at 240V in the British Isles, but some heavy users can receive 415V and industries either 11kV or 33kV.

The local substations, like the electricity pylons, are not playgrounds. Every year young people are killed because they ignore all the warnings and venture in. With the energy supplied to these, very few people get a second chance if they touch the equipment.

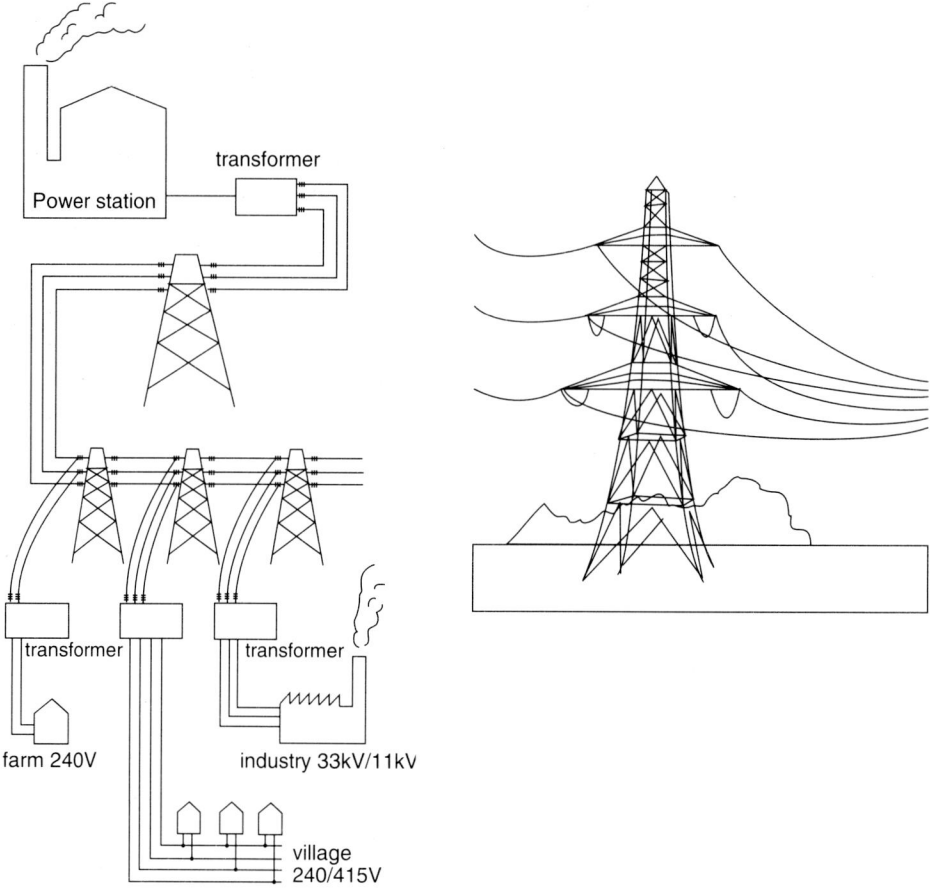

The National Grid

a.c. Supply

The electricity we use from the power sockets in our home and that used by industry is different from the electricity we get from batteries. The electricity from batteries is direct current (d.c.) where the mains supply is alternating current (a.c.). In the United Kingdom, a.c. electricity has a frequency of 50Hz (50 cycles per second). This means that in every second the live a.c. wire will be changing from negative to positive and back 50 times.

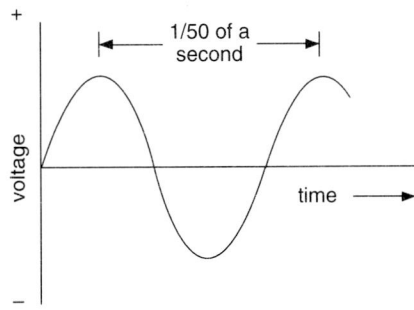

Our homes are supplied with 240V a.c. electricity. What is not generally known is that the electricity arriving at your house is slightly different from that arriving next door. Their electricity is said to be out of phase with yours. The next house along will be slightly later still. The house after that will be in phase with your house. The electricity lines outside the house are termed three-phase power carrying cables. All the houses are connected to the same neutral wire. The voltage difference between your live and your neutral wires is 240V, but between your live and your neighbour's live wire there is a difference of 415V. It is vital to keep appliances powered from two different homes apart, just in case something goes wrong. Remember, 240 volts can kill, so 415 volts is even more dangerous.

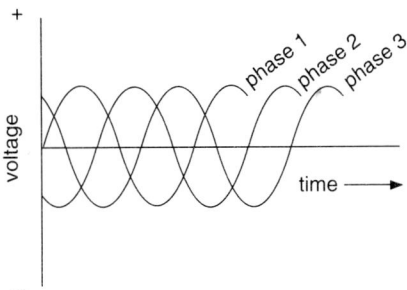

A graph showing the three parts of three-phase electricity. Notice each phase is in a slightly different time to the others.

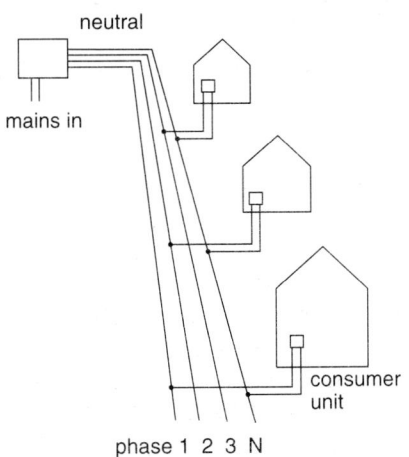

The diagram shows a a three-phase electricity supply along a street; each house's consumer unit is connected to a different phase from that of its neighbour.

Three-phase supply is used to provide electricity for large buildings, such as schools, light industry or tower blocks. Inside the buildings, single-phase 240V supplies are used for ordinary electrical use, with a different phase being used for different floors, and for lighting etc. The boilers in a school or tower block, or for certain high power consumption machines may require a 415V supply.

In a typical tower block, the three-phase supply would be connected to a master cut-out switch. From there it would probably be connected to a 'rising busbar' - four uninsulated metal strips which rise to the top of the building. It is housed safely inside an insulated chamber which prevents the buildings coming into contact with 415V. On each floor, a distribution board, similar to that in your home, is connected up to receive one of the phases of electricity at 240V. If 415V is required for the boiler in the basement the air-conditioning in the roof for example, this can be supplied from the busbar.

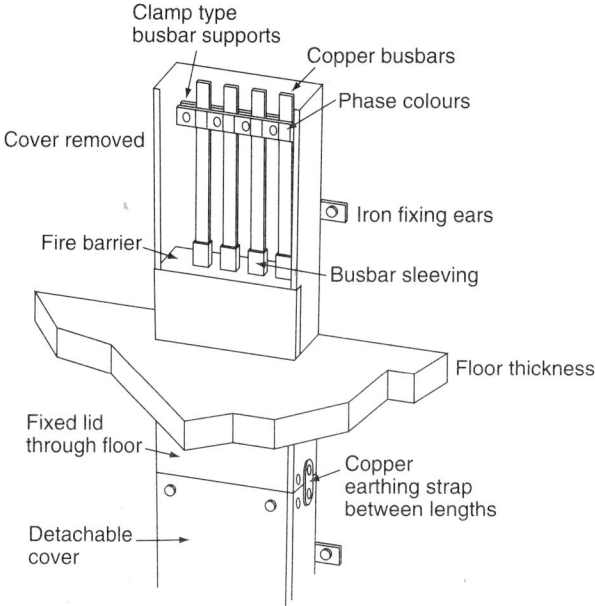

When wiring large areas, the electrician needs to make sure that he does not overload the distribution board. After careful calculation he must decide if one board is enough or whether that particular floor needs two or more distribution boards. In some office blocks and other buildings which have three-phase supply, fluorescent lights can be powered by this supply. Great care has to be taken when fitting these to make sure that the individual live wires are well separated to avoid 415V shock dangers. All electricity supplies, three-phase in particular, must be treated with great care and, unless you know exactly what you are doing, should be left alone.

Electricity into the Home

Electricity usually enters most homes from under the ground. A thick, two-core cable connected to a mains cable in the road comes up through the floor and enters a sealed box. This box contains the electricity company's fuse called the service cut-out and it is illegal to tamper with it.

A two-core cable connects the sealed fuse to the electricity meter which measures how much electricity is used by the household. This in turn is connected to the main fuse box for the house, usually known as the consumer unit, which contains the individual fuses or MCBs to protect the separate circuits in the house. An earthing conductor is usually connected to the earth terminal of the electricity company's sealed fuse box and the main fuse box for the house.

The two wires which carry the electricity into the home are known as the live conductors. The phase is coloured red, and the neutral, which is coloured black. It is the red wire to which the fuses and switches are connected so that they can stop the flow of energy if necessary.

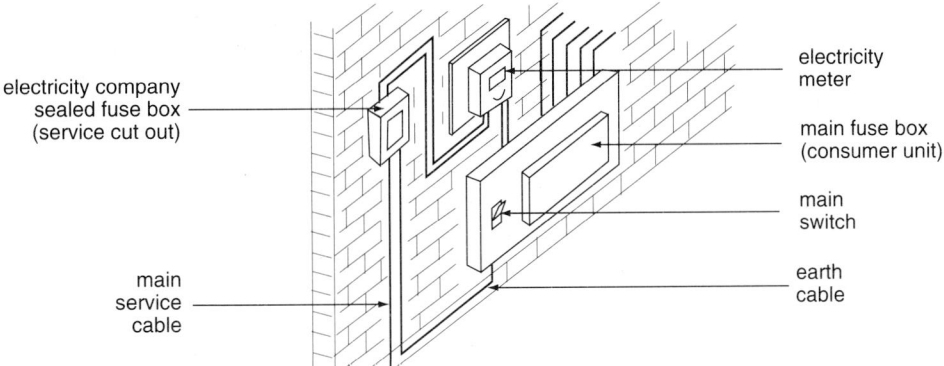

The electricity meter in older wiring systems will have little dials showing how much electricity is used, whereas modern systems will have digital displays. Reading the old type of meter takes some skill as some of the dials move clockwise and others move anti-clockwise. Both types of meter measure the amount of electricity used in kilowatt hours, not in joules as you might expect. One kilowatt hour is the amount of electricity that a 1000 watt appliance (e.g. a kettle) would use if it were left working for one hour. One kilowatt hour is equal to 3,600,000 joules or 3.6MJ (Megajoules).

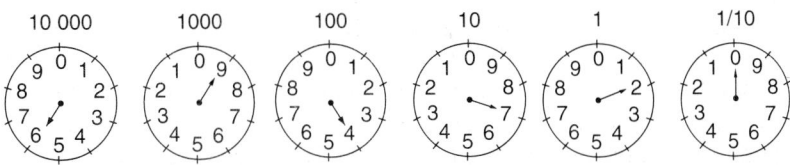

The picture of the dial-type meter shows you how difficult it is to read these meters. The unit dial (read clockwise) is on the 2, so it is read as 2. The ten's dial (read anti-clockwise) is just past the 7 and is read as 7. The hundred's (clockwise) is just before the 5 and so is read as 4 (it has not yet reached 5). The thousand's (anti-clockwise) is half-way between 9 and 0 and so is read as 9. Finally, the ten thousand's (clockwise) is nearly at the 6 and is therefore read as 5. As you can see, it can be confusing, so be careful and check readings at least twice.

The electricity company uses the reading taken by their representative along with the previous reading to calculate how much has been used. They subtract the old reading from the new to show how many kilowatt hours (units) of electricity have been used. They take this figure and multiply it by the cost of one unit to calculate how much money the electricity will cost. This cost is added to a 'standing charge' to produce the bill that has to be paid by the household.

Some houses have two meters, or one with two digital displays. One meter, or display, works for seven hours at night, and the other works for the rest of the day. The night electricity is charged at a much lower rate and is known as 'Economy 7'. This is to encourage the user to use the electricity to heat the house and hot water at night when there is less demand on the power stations.

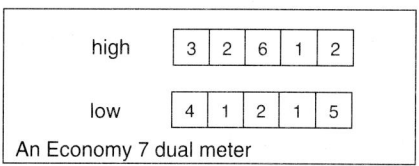
An Economy 7 dual meter

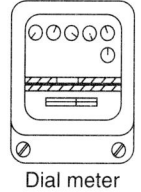
Dial meter

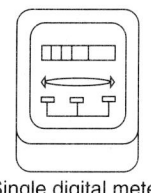
Single digital meter

Extra Low Voltage - (ELV)

Not exceeding 50V a.c. or 120V ripple-free d.c.

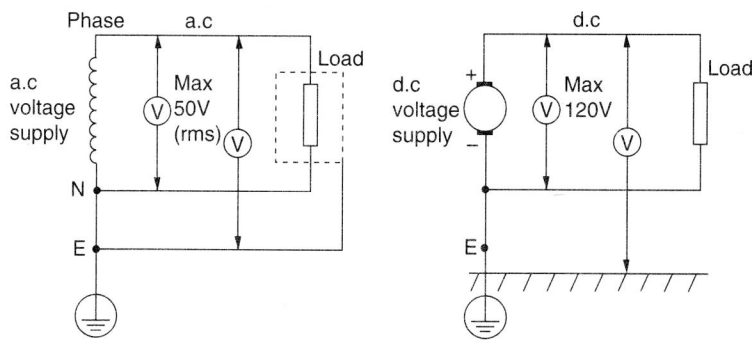

Low Voltage

Exceeding ELV, but not exceeding 1000V a.c. (rms) or 1500V d.c. between conductors or 600V a.c. (rms) or 900V d.c. between conductors and earth.

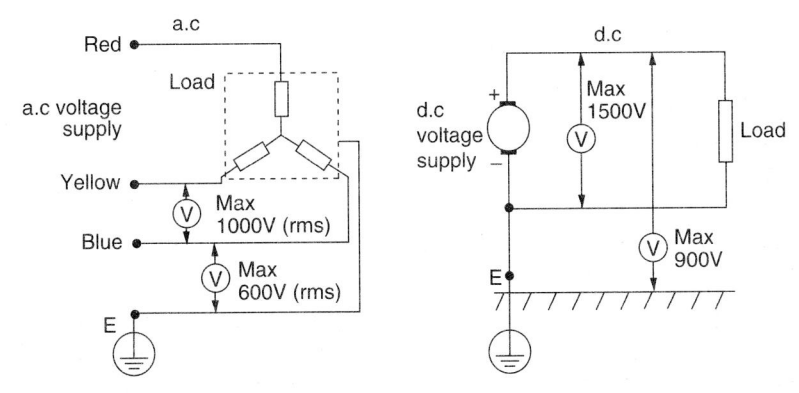

Classification of Systems

A system consists of an electrical installation connected to a supply. Systems are classified with the following letter designations.

SUPPLY earthing arrangements are indicated by the first letter.
T - one or more points of the supply are directly connected to earth.
I - supply system not earthed, or one point earthed through a fault-limiting impedance.

INSTALLATION earthing arrangements are indicated by the second letter.
T - exposed conductive parts connected directly to earth.
N - exposed conductive parts connected directly to the earthed point of the source of the electrical supply.

The **EARTHED SUPPLY CONDUCTOR** arrangement is indicated by the third letter.

S - separate neutral and protective conductors.
C - neutral and protective conductors combined in a single conductor.

The common types of systems are:

TN-S   TT   TN-C-S

System Earthing Arrangements

TN-S Systems

This is likely to be the type of system used where the electricity company's installation is fed from underground cables with metal sheaths and armour. In TN-S systems the consumer's earthing terminal is connected by the supply authority to their protective conductor (i.e. the metal sheath and armour of the underground cable network) which provides a continuous path back to the star point of the supply transformer, which is effectively connected to earth.

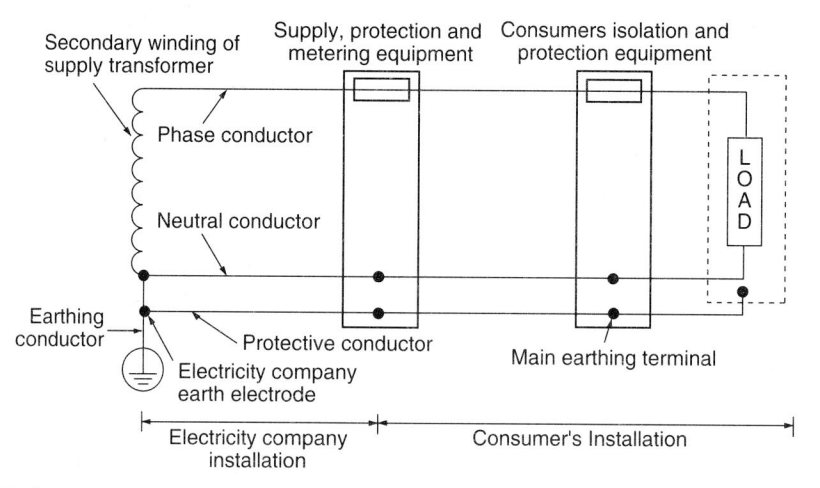

TT Systems

This is likely to be the installation used where the electricity company's installation is fed from overhead cables, where no earth terminal is supplied. With such systems the earth electrode for connecting the circuit protective conductors to earth has to be provided by the consumer. Effective earth connection is sometimes difficult to obtain and in such cases a residual current device should be installed.

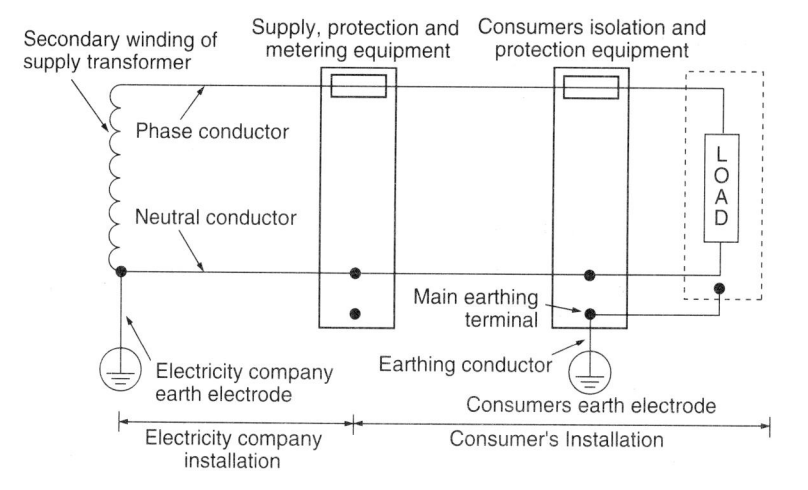

TN-C-S Systems

When the electricity company's installation uses a combined protective and neutral (PEN) conductor, this is known as a TN-C supply system. Where consumer's installations consisting of separate neutral and earth (TN-S) are connected to the TN-C supply system, the combination is called a TN-C-S system. This is the system usually provided to the majority of new installations, referred to as a PME system by the electricity company.

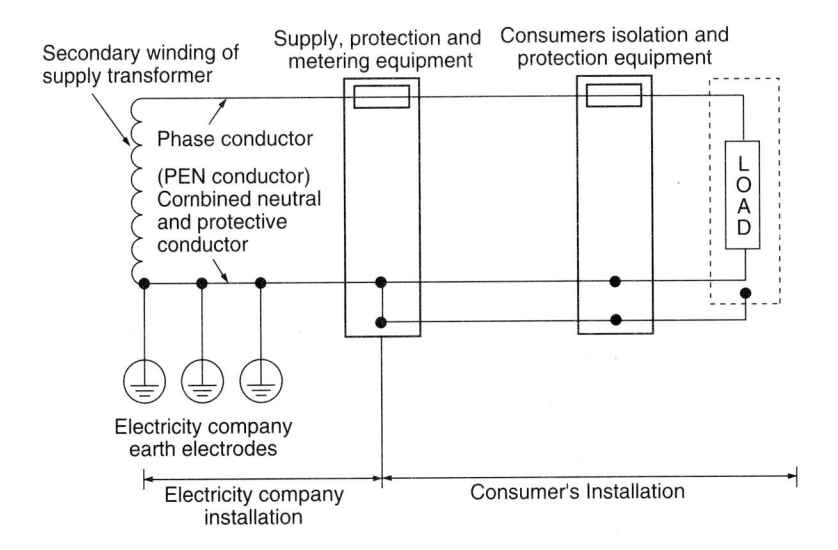

Earthing Arrangements and Terminations

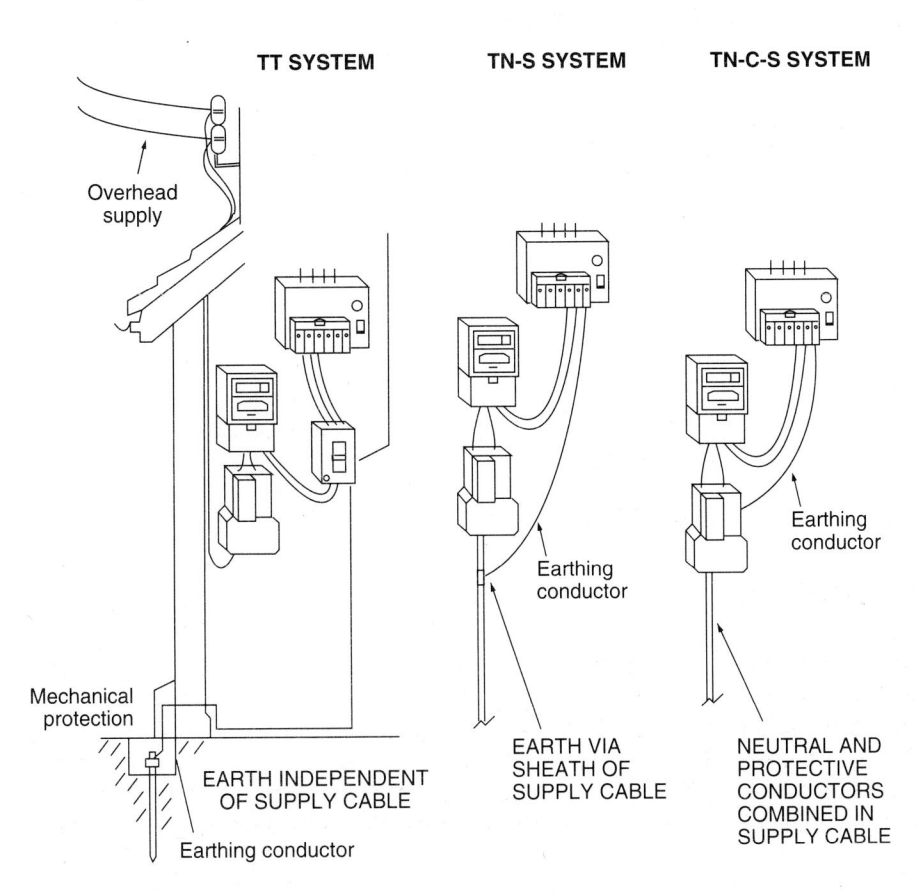

SELV

An extra-low voltage system electrically separated from earth and from other systems in such a way that any fault does not provide risk of electric shock.

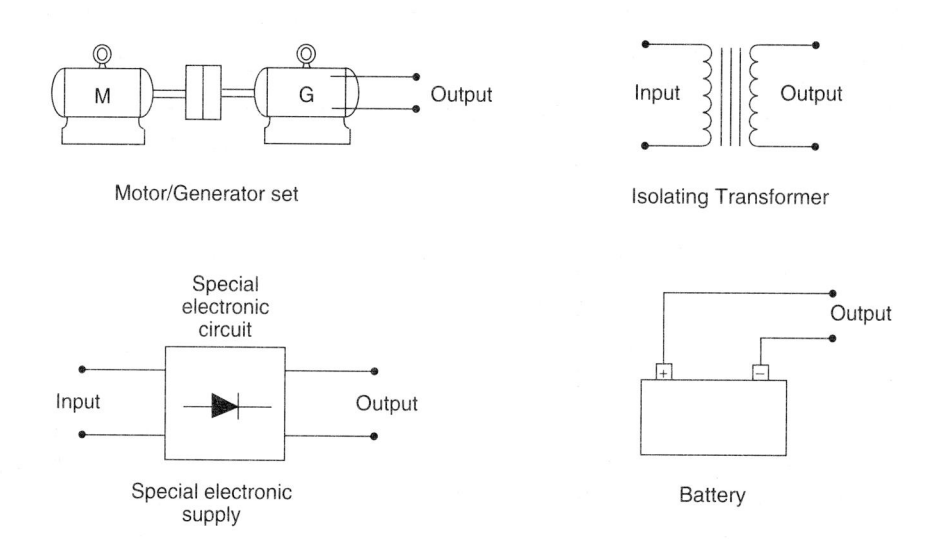

Motor/Generator set

Isolating Transformer

Special electronic supply

Battery

Functional Extra-Low Voltage

Any extra-low voltage system where all of the protective measures required for SELV have not been applied e.g. where the output installation is connected to earth.

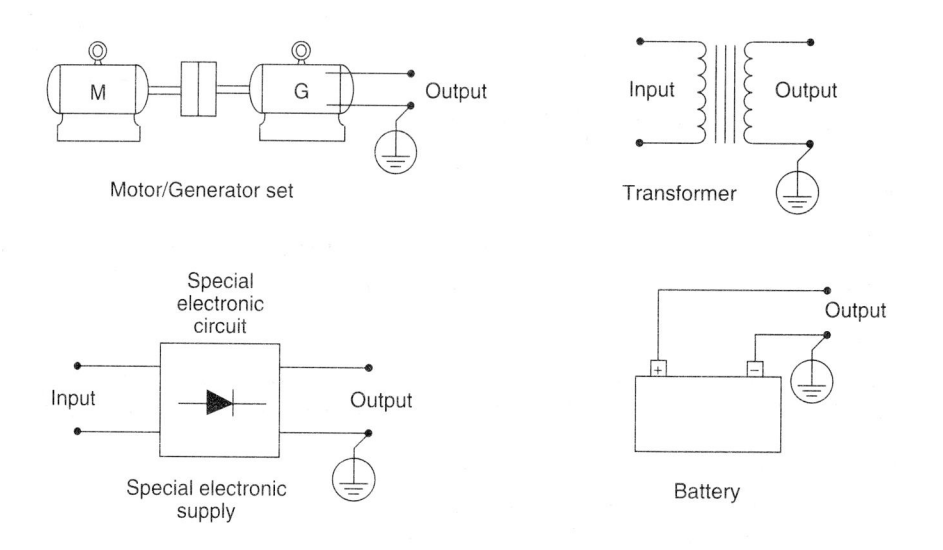

Motor/Generator set

Transformer

Special electronic supply

Battery

Control and Protection Equipment

The Electricity Companies Equipment

The Electricity Generating Companies are responsible for the generation of electricity, but the responsibility for its distribution to consumers is shared by the Regional Electricity Companies. Consumers' installations are connected to the Electricity Companies cables by means of service cables, and these are terminated within the premises at a mutually agreed position.

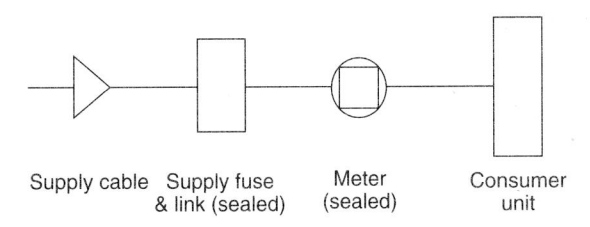

Supply cable   Supply fuse   Meter   Consumer
               & link (sealed)   (sealed)   unit

At the service termination point the service cable is connected to a service fuse. This fuse, which normally has a rating of 80 or 100A, will ensure that no damage occurs to the cables or equipment in the event of a fault occurring on the consumer's premises. Connections are installed by the Electricity Companies from the service fuse to the meter, which is supplied to record the amount of electricity consumed. The service fuse and meter are sealed by the Electricity Company and consumers or contractors are not permitted to interfere with this part of the installation.

The Consumer's Installation

Connections are made from the meter to one or more main switches, and it is at this point that the consumer's responsibility for the installation begins. Main switches must be provided so that the installation may be isolated from the supply when alterations or extensions are made.

Consumer's Control Unit

For domestic installations where the load does not exceed 100 amperes the use of a consumer's control unit is recommended.

Consumer's control units are made to British Standards which usually consist of a 80/100 ampere double pole switch and a number of fused outlets. An arrangement with eight fuse ways may be used to supply the following circuits.
Two 30 ampere ways for two ring final circuits
One 30 ampere way for the cooker circuit
One 30 ampere way for the electric shower
One 15 ampere way for the immersion heater circuit
Two 5 ampere ways for the lighting circuits
One spare way

There will be many variations; no two types of installation will have exactly the same needs.

Each consumer unit will have its own double pole main switch, fuseholders of miniature circuit breakers for each way, neutral terminals and earth terminals.

Metal clad or insulated enclosures may be installed to accommodate the consumer unit. If a conduit system is being installed a metal clad fuseboard will be used. For sheathed cable systems an insulated enclosure is more usual.

Main Switchgear

Every installation must be controlled by one or more main switches. The main switchgear may consist of a switchfuse or a separate switch and fuses and must be readily accessible to the consumer and as near as possible to the Electricity Companies service intake. The requirements for switchgear use for protection, isolation and switching are given in Chapter 55 of the IEE Regulations.

The general requirements for isolation and switching in Chapter 46 of the IEE Regulations states that *'every installation and all items of equipment should be provided with effective means to cut off all sources of voltage.'*

Main switchgear. Permissible combinations

Control of Separate Installations

Control apparatus must be provided for every section of a consumer's installation. An off-peak supply, such as that supplying electric storage heaters, is considered to require separate metering and control apparatus in addition to the metering and control apparatus for lighting and other loads. Where a consumer's installation is split up into separately controlled parts, each part must be treated as a separate installation. This applies irrespective of whether the parts are within the same building or in separate detached buildings.

Every installation must be controlled by one or more main switches. The main switchgear may consist of a switchfuse or separate switch and fuses.

Control of detached buildings

Installations in detached buildings must each be provided with a specified means of isolation for example as illustrated.

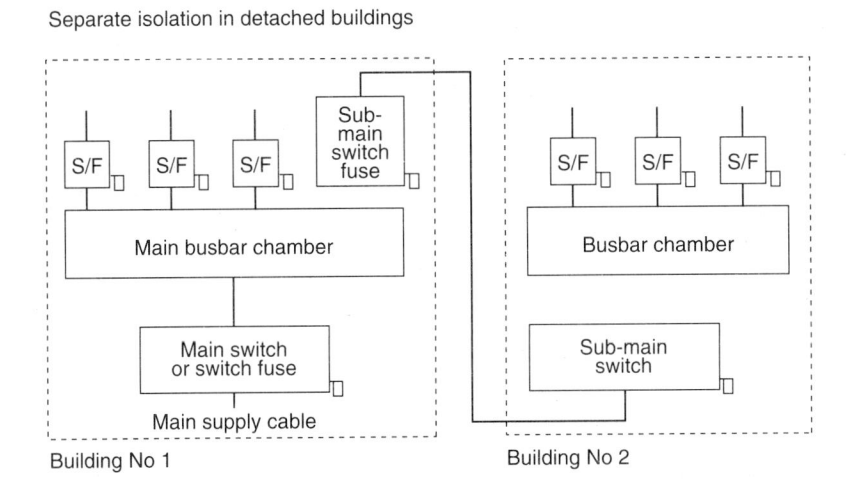

Separate isolation in detached buildings

Building No 1

Building No 2

Identification Notices

Switchgear Control Gear and Protective Devices

Switchgear and control gear in an installation should be labelled to indicate its use. All protective devices in an installation should be arranged and identified so that their respective circuit may be easily recognised.

Diagrams and charts must be provided for every electrical installation indicating:

(a) the type of circuits
(b) the number of points installed
(c) the number and size of conductors
(d) the type of wiring system
(e) the location and types of protective devices and isolation and switching devices

Note: For simple installations the foregoing information may be given in a schedule, if symbols are used they should conform to BS 3939.

The purpose of providing diagrams, charts and tables for an installation is so that it can be inspected and tested in accordance with Part 7 of the Regulations to provide any new owner of the premises (should the property change hands) with the fullest possible information concerning the electrical installation.

It is essential that diagrams, charts and tables are kept up to date.

A typical chart for a small installation is illustrated:

Type of circuit	Points served	Phase Conductor mm^2	Protective Conductor mm^2	Protective devices	Type of wiring
Lighting	10 downstairs	$1mm^2$	$1mm^2$	5 Amp Type 2 MCB	PVC/PVC
Lighting	8 upstairs	$1mm^2$	$1mm^2$	5 Amp Type 2 MCB	PVC/PVC
Immersion Heater	Landing	$2.5mm^2$	$1.5mm^2$	15 Amp Type 2 MCB	PVC/PVC
Ring	10 downstairs	$2.5mm^2$	$1.5mm^2$	30 Amp Type 2 MCB	PVC/PVC
Ring	8 downstairs	$2.5mm^2$	$1.5mm^2$	30 Amp Type 2 MCB	PVC/PVC
Shower	Bathroom	$6mm^2$	$2.5mm^2$	30 Amp Type 2 MCB	PVC/PVC

BS 4363 Distribution Units

Distribution units consist of earth-proof cabinet enclosures designed to be freestanding or wall mounted. The normal method of connection is by plug/socket appliance inlet-coupler. In this way, both input and output supply connections can be easily and safely made by unskilled personnel. Permanent connections using glands are also available. Plugs and sockets can either be splashproof or watertight.

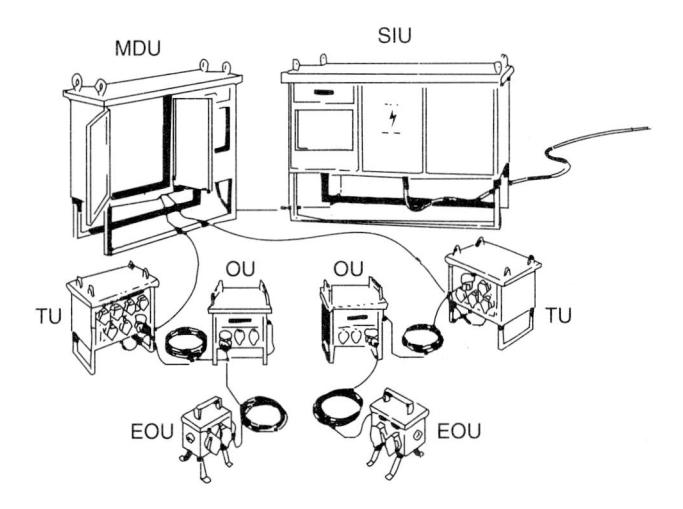

Each unit of equipment is identified by an abbreviation
- SIU - supply incoming unit
- MDU - main distribution unit
- SIDU - supply incoming and distribution unit
- TU1/3 - transformer unit
- OU/1 - (single phase) outlet unit
- OU/3 - (three phase) outlet unit
- EOU/1 - (single phase) extension outlet unit
- EOU/3 - (three phase) extension outlet unit.

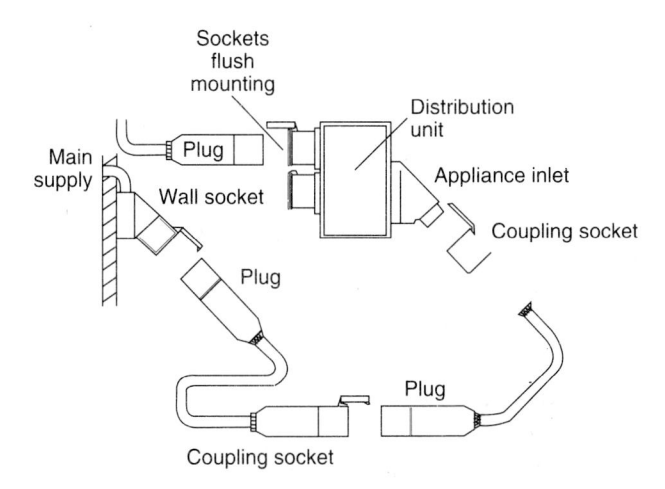

Protection of both input and output supplies is normally by miniature circuit breakers for supplies up to 70A and by moulded case circuit breakers for capacities exceeding this level. If preferred, HBC fuse-switch protection can be substituted.

BS 4343 Plugs and Sockets

The range of accessories consists of:

- plugs
- sockets
- cable couplers
- appliance inlets

Accessories are available for single and three-phase supplies with a voltage between phases not exceeding 750V at a rated current of up to 125A. Discrimination between different voltages is achieved in two ways:

- by colour code
- by the positioning of the earth contact in relation to a keyway.

Colour code and earth contact relationship to keyway chart are as illustrated below. References should be made to current manufacturer's literature.

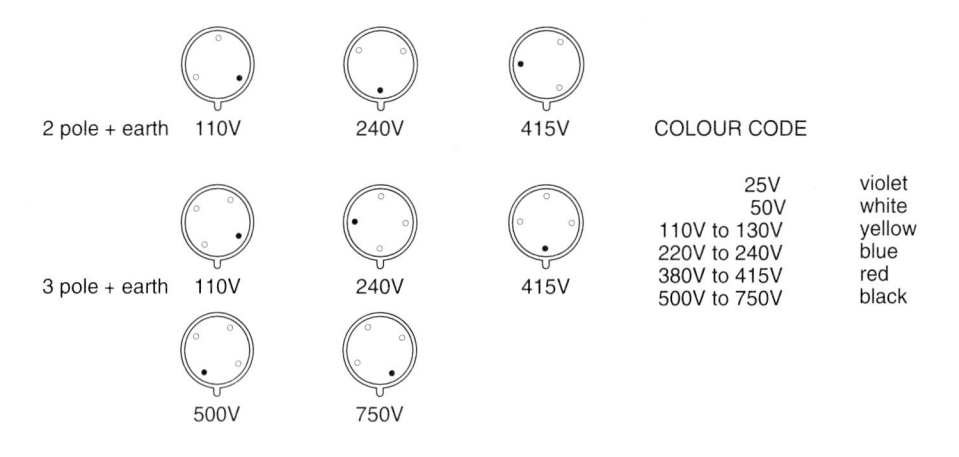

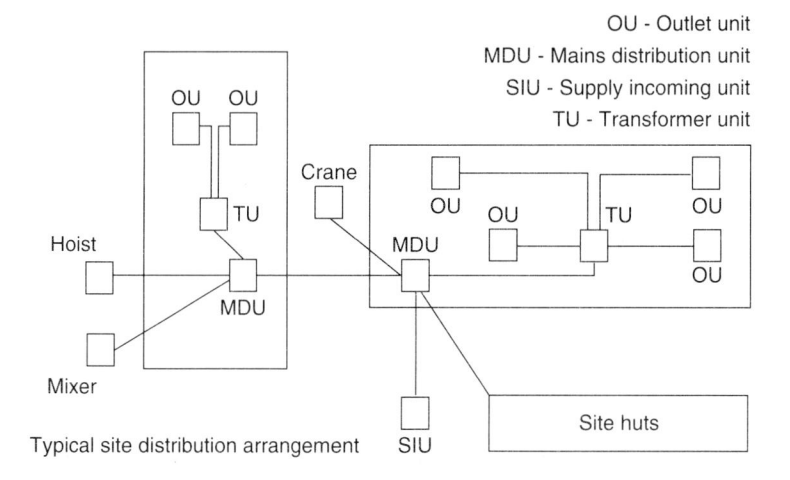

2 pole + earth   110V   240V   415V   COLOUR CODE

25V	violet
50V	white
110V to 130V	yellow
220V to 240V	blue
380V to 415V	red
500V to 750V	black

3 pole + earth   110V   240V   415V

500V   750V

OU - Outlet unit
MDU - Mains distribution unit
SIU - Supply incoming unit
TU - Transformer unit

Typical site distribution arrangement

Typical Rating of BS 4363 Distribution Equipment

Symbol	Outputs			Inputs		Power output at 110v	
	2 pole + E	3 pole + E	2/3 pole + E	V	A	kVA	A
TU/1	●			240	5	1	9
					10	2	18
					20	5	45
					45	10	90
					60	15	135
TU/3		●		415	5	5	25
					20	15	75
TU/1/3			●		35	25	125
OU/1	●			110	30	3	30
OU/3		●		110	25	5	25
EOU/1	●			110	15	1.5	15
EOU/3		●		110	15	3	15

Calculation of Single Phase Loads

Voltage x Current = VA
1,000 VA = 1kVA

Therefore if the voltage of a system is known and the kVA rating of a distribution unit is known then the maximum load current that unit can supply can be calculated by:

$$\frac{kVA \times 1{,}000}{voltage} = current\ (Amps)$$

A typical example of calculating the load of a site distribution system is as illustrated.

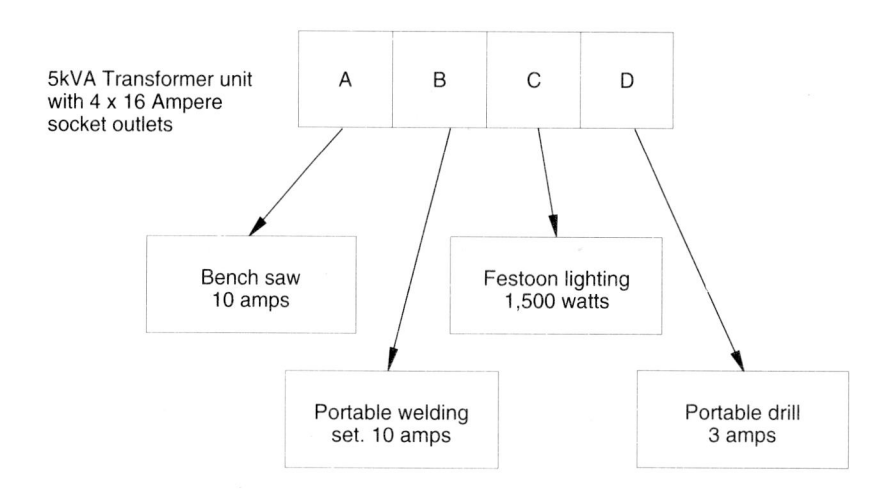

```
5kVA Transformer unit     ┌───┬───┬───┬───┐
with 4 x 16 Ampere        │ A │ B │ C │ D │
socket outlets            └───┴───┴───┴───┘
```

- A → Bench saw 10 amps
- B → Portable welding set. 10 amps
- C → Festoon lighting 1,500 watts
- D → Portable drill 3 amps

Transformer unit
maximum load = $\dfrac{5{,}000\text{VA}}{110\text{V}}$ = 45 Amps

Lighting load

= $\dfrac{\text{Watts}}{\text{Voltage}} = \dfrac{1{,}500\text{W}}{110\text{V}}$

= 14 Amps

Total load = A + B + C + D

= 10 + 10 + 14 + 3
= <u>37 Amps</u>

Responsibility and Testing of Construction Site Installations

The installation of an electricity supply to a construction site should preferably be in the charge of a competent person who has full responsibility for the safety and use of the installation and for any alteration or extension of the circuit.

It is essential that sub-contractors and others who bring electrical equipment onto a construction site should first consult with the person responsible as to the suitability and safety of the equipment, that the correct type of plugs are fitted and that the voltage and load required to operate the equipment can be supplied safely.

Instrumentation

Measurement of Voltage, Current and Resistance

Instrument and Display Modes

The most common type of instrument is the multi-test meter, which is a meter capable of measuring values of voltage, current and resistance.

You will meet many different makes and models in your work but they usually fall into one of two categories, analogue or digital.

The analogue meter uses a moving needle against a scale, which has to be interpreted, like a watch with hands. The digital meter displays a read out in numbers, the same as a digital watch.

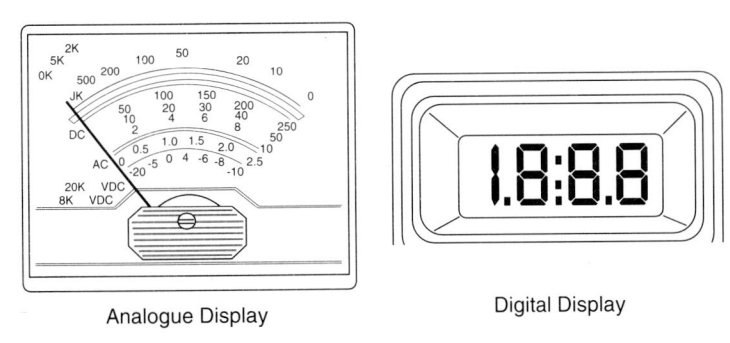

Analogue Display

Digital Display

Selection of Correct Function and Range

By turning the selector switch you are able to select the electrical unit you wish to measure, and the range.

        DC Voltage   AC Voltage
        DC Current   AC Current

        and resistance in OHMS

Remember whenever you have finished using the meter always turn the selector switch to the off position.

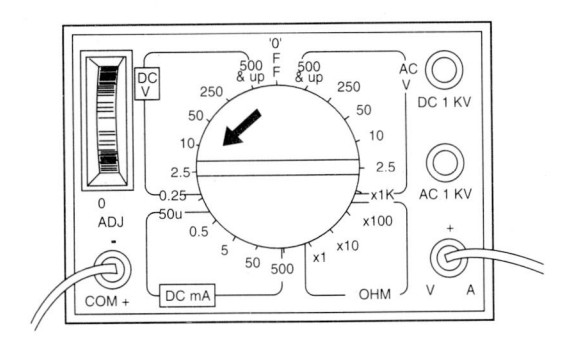

3/1

Calibration

Before the instrument is ready for use. The instrument should be checked for correct calibration. The state of the battery and continuity of the test leads should also be checked. Start by fitting the test leads in the correct terminals red +, black -.

Correct Calibration

Set the range switch to the lower value of d.c. voltage and touch the test leads together, the needle should read zero, or the digital read out should give zero. If the value is not zero the accuracy of readings will be affected. In order to adjust the reading to zero the adjustment screw must be turned until zero is obtained.

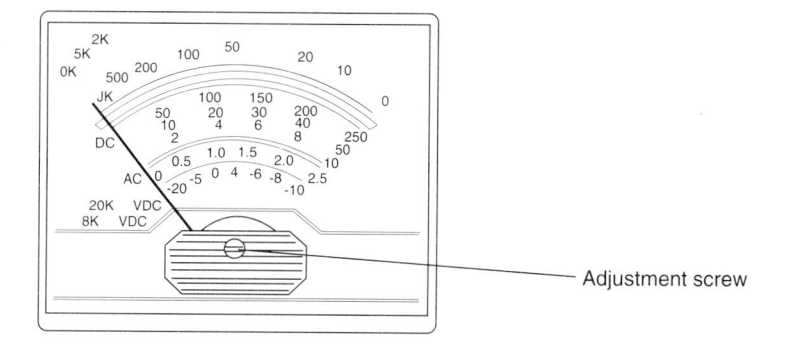

— Adjustment screw

Battery Condition and Continuity of Test Leads

To check, select the lower ohm's range and bring the test probes into contact with each other. If the battery is healthy and the test leads are satisfactory you will be able to zero the reading on the ohm's scale using the adjuster. If there is no movement of the pointer, it is likely that the test leads or connections are faulty. If there is movement but you cannot zero the pointer then the battery is suspect.

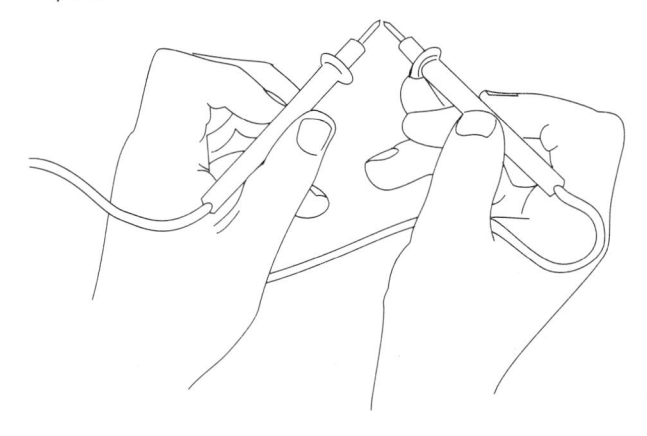

Remember there are things that have to be carried out every time you use the meter.

The meter must be calibrated.

The condition of the battery and continuity of the test leads must be checked.

The meter must be in the correct operating position, usually flat on its back. When the tests are complete turn the selector switch to off.

To measure values of voltage, current and resistance, always select the correct range. If you are unsure then start at the highest range and change progressively to a lower range to achieve a reasonable reading and prevent overloading.

Measuring Voltage

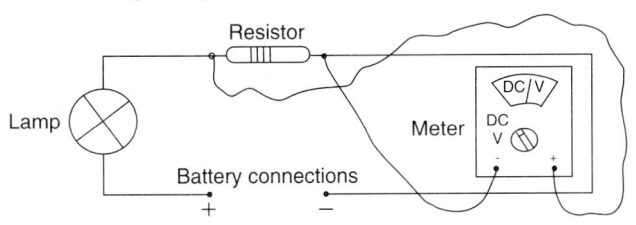

Measuring Current

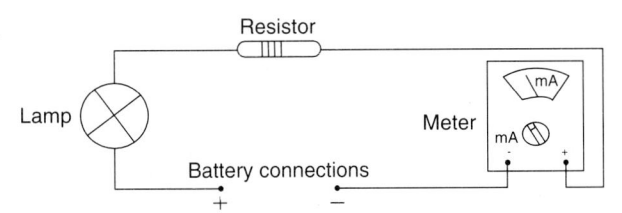

Measuring Resistance

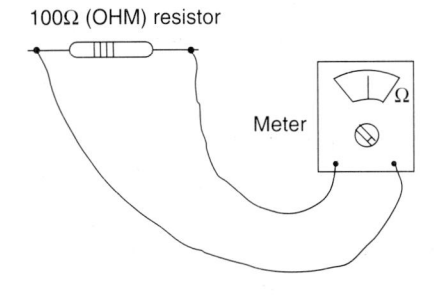

100Ω (OHM) resistor

Remember

When using any meter to measure resistance remember the following in order to obtain accurate readings and to prevent damage to the meter.
Never connect the meter into a circuit with voltage present or current flowing. Always make sure the component or circuit you are checking is isolated from any other components or circuit.

Avoiding Read Errors

Avoid errors on analogue meters which are caused by incorrect position of the meter in relation to your eyes. This error is called parallax error. To avoid parallax error you should ensure all readings are taken with the eye directly above the pointer. If a mirror is fitted to the scale, the pointer's reflection should disappear.

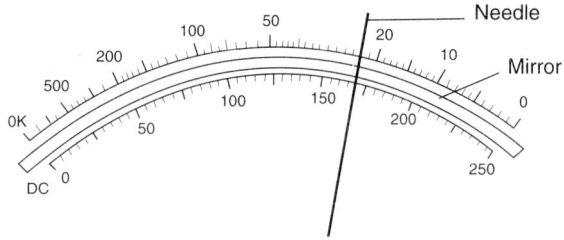

Rules to Follow

The rules for using a multirange meter are as follows:

Place it in its correct operating position

Connect test leads correctly

Calibrate meter

Check the condition of the internal batteries and continuity of the test leads

Adjust zero ohms setting whenever you select a different ohms range

Switch the meter off after use

When using any meter you must never:

Exceed the voltage or current range selected

Subject the meter to reverse polarity

Try to measure the voltage or current when the ohms range is selected

Try to measure any electrical unit with the meter incorrectly set or connected

Try to measure any electrical unit that the meter is not designed to measure

When commissioning electrical installations it is often necessary to check the actual load of individual circuits to verify the design values. To do this would involve disconnecting the circuits and inserting the ammeter test leads in series with the loads to determine the current drawn from the supply. An alternative and much simpler method is to use a clamp type instrument as shown blow. This uses the current transformer principle to enable the current to be measured without disconnecting the circuit conductors.

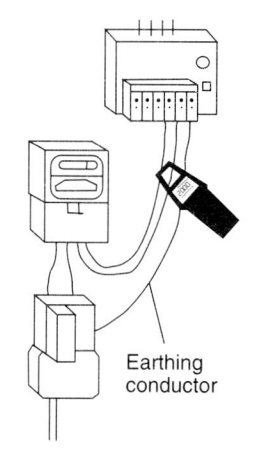

Measuring Resistance of Electrical Installations

Earth Continuity Testing

When carrying out electrical installation work and maintenance, it is often necessary to verify that a conductor is continuous and not broken. This is called continuity testing and involves the use of a low resistance reading ohmmeter, capable of measuring resistance values of less than 5Ω.

The situation where this type of testing is frequently required is after the installation of a circuit protective to ensure the earth wire is both continuous and effective. Another situation would be in kitchens and bathrooms where the plumbing and heating installation is simultaneously accessible with the metal work of electrical appliances or equipment and effective bonding of taps and sink top is required.

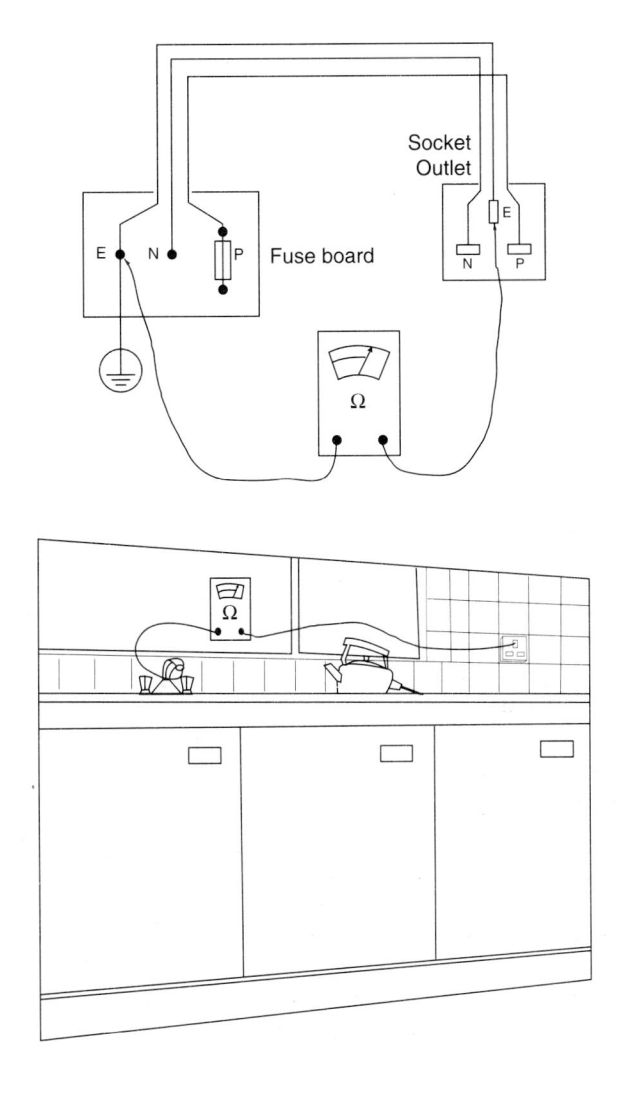

Insulation Resistance Testing

In order to ensure an electrical installation or equipment which has been worked on is safe to leave without the risk of electric shock, possibly caused by a trapped or bare conductor, a test should be made using an insulation resistance tester as follows:

Phase to Earth
Neutral to Earth
Phase to Neutral (except if circuit equipment has electronic components fitted e.g. LCD or LED displays.

Because insulation of conductors has a very high resistance, an ohmmeter or multimeter is unsuitable for carrying out the test as it is incapable of supplying sufficient test voltage. Therefore an insulation resistance tester capable of delivering 500 volts d.c. with a load current of 1mA should be used as indicated. The value of resistance in MΩ must be greater than 0.5MΩ if the circuit or equipment is safe to connect to the supply as stated in the IEE Wiring Regulations.

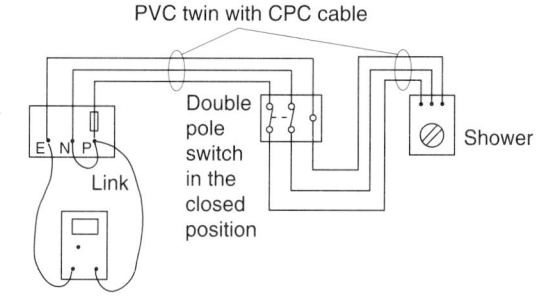

Avoiding Errors

Before using an insulation resistance and continuity tester, always carry out the following checks:

*Battery Condition

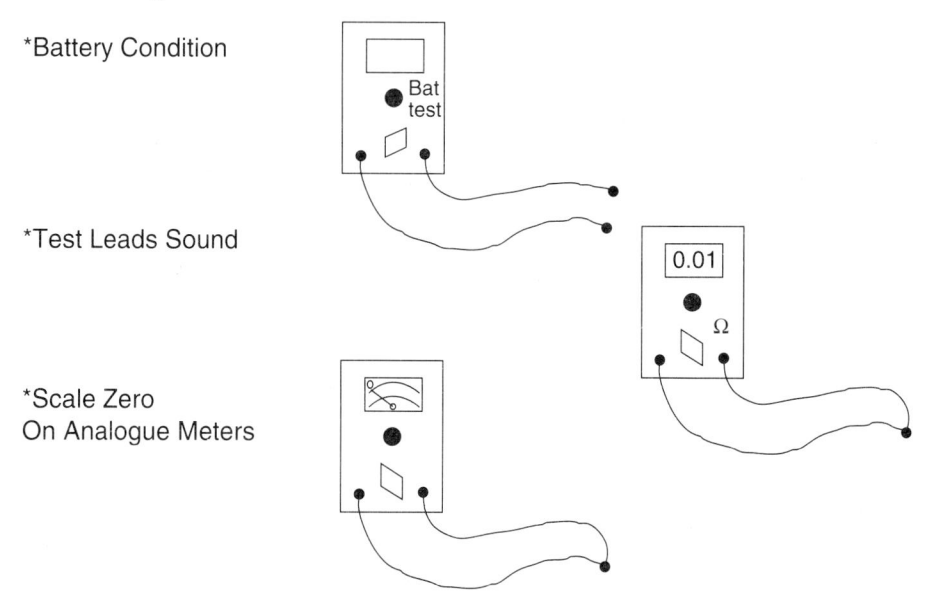

*Test Leads Sound

*Scale Zero
On Analogue Meters

Regulations and Standards

The regulations and standards which cover electrical installations and practices associated with plumbing, gas, heating & ventilating and refrigeration installations can be broken down into categories as indicated.

Statutory

Health and Safety at Work Act 1974
Electricity Supply Regulations 1988
The Low Voltage Electrical Equipment (Safety) Regulations 1989
Electricity at Work Regulations 1989
Building Standards (Scotland) Regulations 1990

British Standard and Codes of Practice

BS 7671 Requirements for Electrical Installations (The IEE Wiring Regulations)
Reference should be made to:

Appendix 1 of BS 7671
Appendix 2 Memorandum of Guidance on the Electricity at Work Regulations 1989

HSE Guidance Notes

Reference should be made to:

Appendix 1 Memorandum of Guidance on the Electricity at Work Regulations 1989

The Health and Safety at Work Act 1974

This act covers all workplaces and is intended to secure and promote health and safety at work. Among other matters it defines the duties of employees. All employees should know what their responsibilities are.

The act requires you to:

Exercise reasonable care for your own health and safety and that of others who might be affected by what you do or fail to do.

Co-operate with your employer to enable him to comply with the requirements of the act.

Not interfere with or misuse anything provided in the interests of health and safety.

The act also imposes duties on employers and employees to:
Ensure plant and machinery are safe and that safe systems of work are set and followed.

The Electricity Supply Regulations 1988

These regulations impose requirements with regard to the installation and use of electric lines and apparatus of suppliers of electricity including the provisions for connections with earth.

The regulations also specify that the voltage at the supply terminals shall be no greater than 6% above or below the declared voltage.

The Low Voltage Electrical Equipment (Safety) Regulations 1987

These regulations apply to a.c. equipment operating at voltages above 50 volts and below 1,000 volts. They also apply to d.c. equipment operating above 75 volts and below 1500 volts.

General Conditions

The electrical equipment together with its component parts should be made in such a way to ensure that it can be safely and properly assembled and constructed.

The equipment should be designed and manufactured so as to ensure protection against hazard providing that equipment is used in application for which it was made and is adequately maintained.

Protection from Hazards

Persons and domestic animals shall be protected from danger or physical injury or other harm which might be caused from direct or indirect contact.

Electricity at Work Regulations

Introduction

These regulations came into force on the 1st April 1990. They have been written to reduce the increasing number of accidents involving electricity.

Their purpose is to require precautions to be taken against the risk of death or injury from electricity in work-related activities. The emphasis is on prevention of danger from electric shock, burns, electrical explosion or arcing or from fire or explosion initiated by electrical energy.

The regulations apply wherever the Health and Safety at Work Act applies, wherever electricity may be encountered and to all persons at work. Areas covered are those associated with the generation, provision transmission, transformation, rectification, conversion, conduction, distribution, control, storage, measurement or use of electrical energy, for example, from a 400kv overhead line to a battery--powered torch. Therefore there are no voltage limits to these regulations.

In order to provide guidance on the interpretation of the regulations the Health and Safety Executive (HSE) have produced a memorandum of guidance to assist persons involved in the design, construction, operation or maintenance of electrical systems and equipment. Therefore the assessment of danger and how the regulations are to be applied to overcome it, will be the continuous responsibility of both the employer and employee working as a team to achieve continual compliance with the regulations.

When persons who design, construct, operate or maintain electrical installations and equipment need advice they should refer to guidance, such as may be found in codes of practice or HSE guidance, or they should seek expert advice from persons who have the knowledge and experience to make the right judgments and decisions; the necessary skill and ability to carry them into effect. It must be remembered a little knowledge is usually sufficient to make electrical equipment function but it usually requires a much higher level of knowledge and experience to ensure safety.

The Regulations

The following is a brief summary of each of the regulations with the emphasis on electrical systems and equipment associated with plumbing, heating & ventilating, refrigeration and gas installations.

Regulation 1 Citation and Commencement

The regulations are entitled The Electricity at Work Regulations 1989 and came into force on 1st April 1990.

Regulation 2 Interpretation

Explains the meaning of the following:

Systems
Electrical Equipment
Conductors
Danger
Injury

Regulation 3 Persons on Whom Duties are Imposed by These Regulations

It shall be the duty of every employer and self employed persons to comply with the regulations for matters within his control.

It shall be the duty of every employee when at work to co-operate with his employer to enable him to comply with the provisions of the regulations and also to comply with the regulations for matters within his control.

Regulation 4 Systems, Work Activities and Protective Equipment

This regulation has very wide application and requires all electrical systems to be constructed and maintained in a way so as to prevent danger. This includes the design of the system and correct selection of equipment, as well as regular inspection and maintenance to ensure the continuing safety of the system and associated equipment.

The requirement for maintaining detailed and up-to-date records is also stressed.

System of Construction

It is the responsibility of the Designer, Installer and Tester to meet this obligation.

The electrical test is carried out on the commissioning of the installation for the purpose of:

Verifying the theoretical design intent against actual instrument test values.

System Maintenance

It is the responsibility of the Owner or Duty Holder to meet this obligation.

This refers to the necessity to monitor the condition of the system throughout its life, by implementing a programme of periodic testing and inspecting.

Records of maintenance, including test results should be kept, thus monitoring the condition of the installation, and the effectiveness of the programme.

With regard to work activities the overriding consideration is that work should not be carried out on a system unless it is "dead". The circuit or equipment to be worked on must be properly isolated by locking-off, labelling etc. and the circuit proved dead at the point of work before work starts.

Systems with more than one possible supply e.g. extractor fans with timer circuits will need extra care to ensure all relevant circuits are properly isolated. The test instrument used e.g. multimeter or approved test lamp to prove the circuit or equipment dead, must itself be proved as functioning correctly, immediately before and after testing. This can be achieved using a known supply point or a voltage proving unit.

The last part of this regulation includes the requirement that protective equipment such as special tools, protective clothing and insulating materials, must be suitable for the purpose, maintained in good condition and be properly used. Examples are insulating gloves and floor mats which are covered by British Standards.

e.g. BS 697 Specification for Rubber Gloves for Electrical Purposes
BS 921 Specification for Rubber Mats for Electrical Purposes.

Regulation 5 Strength and Capabilities of Electrical Equipment

All equipment must be selected so that it meets appropriate standards e.g. British Standard, operates safely under normal and fault conditions, in order to ensure its ability to withstand thermal, electro-magnetic, electro-chemical or other effects of electrical current which will flow when a system is operating. Equipment must be installed and used in accordance with the instructions supplied.

Regulation 6 Adverse or Hazardous Environments

This regulation covers the requirement to consider the kinds of adverse conditions where danger would arise if equipment is not constructed and protected to withstand such exposure. Therefore it will be necessary at the design stage or when considering maintenance requirements of equipment to consider the effects of mechanical damage, weather and corrosive substances.

Examples of protection from mechanical damage would be running cables through holes in the centre of a joist instead of in a slot at the top of the joist where the cable could be struck by a nail when fastening down the floorboards.

In the case of weather, when installing frost thermostats the thermostat would have to be suitable for outside use with a suitable enclosure (case) and the cable entry would need to be either sealed around the cable using a suitable compound or the termination made using a weatherproof gland.

Regulation 7 Insulation, Protection and Placing of Conductors

This regulation requires that all conductors in a system which may give rise to danger to be covered with suitable insulating material.

The IEE Wiring Regulations provide information and advice for electrical installations up to 1,000 volts only.

Regulation 8 Earthing of Other suitable Precautions

This regulation applies to any conductor, other than a circuit conductor, which may become charged with electricity either as a result of use or a fault occurring in the system, e.g. circuit protective conductes (c.p.c's)

Typical techniques are:

- Double insulation (portable tools)
- Earthing to a mass of earth (A system used when the electricity company does not provide an earth)
- Equipotential bonding (The system most commonly used in all electrical installations)

- Use of safe voltages (110 volt transformers max voltage 55 volts to earth)
- Use of RCD's (Provide additional protection to fuse & MCB's)
- Separated or isolating transformers (found in bathroom shaver sockets and the supply system for whilpool baths).
- Double- wound transformers (As used in 110V transformers)

Regulation 9 Integrity of Referenced Conductors

This refers to the maintenance of the integrity of the earth and/or neutral conductors, for example, not inserting fuses, MCB's or switches into conductors connected to earth.

Regulation 10 Connections

This regulation requires all joints and connections to be both mechanically and electrically suitable for their use.

It therefore covers connections to terminals of plugs, socket outlets, fused spur units, junction boxes and appliances. Therefore all terminations of cables and flexible cords shall be such that the correct amount of insulation is removed, the connections are tight and verified by a tug on the conductor and that no stress or strain is placed on the conductor or cable itself which can be achieved by using cord grips in plug tops or appliances and by clipping cables around entries into junction boxes.

Regulation 11 Means of Protection from Excess of Current

To meet the requirements of this regulation, the means of protection is likely to be fuses or circuit breakers to guard against overload, short circuit and earth fault current, the IEE Regulations provide guidance on this subject. However, the type of electrical protection must be properly chosen, installed and maintained in accordance with good practice.

Regulation 12 Means for Cutting Off the Supply

This regulation covers two separate functions. Cutting-off the supply and isolation.

Devices used for cutting-off the supply are usually switches which must be capable of disconnecting the supply to equipment under normal operating or fault conditions.

They should be clearly marked to show their relationship to the equipment they control unless that function is obvious to persons who need to operate them.

Devices used for isolation are usually switches which have the capability of establishing, when operated, an air gap with sufficient clearance distances to ensure that there is no way in which the isolation gap can fail electrically. The position of the contacts or other means of isolation should either be externally visible or clearly and reliably indicated.

Typical isolating devices would be a double-pole switch in a consumer unit and in the case of an appliance a 13 amp plug and socket.

Both devices for switching and isolating should be positioned so that there is ease of access and operation and the area adjacent is kept free from obstructions.

Regulation 13 Precautions for Work on Equipment Made Dead

This regulation states the requirements for such precautions should be effective in preventing electrical equipment from becoming charged with electricity in such a way that would given rise to danger.

The procedures for making equipment dead will usually involve switching off the isolating device and locking it off. If such facilities are not available, the removal of fuses or links which have to be kept in safe keeping can provide an alternate and secure arrangement if proper control procedures are used.

Regulation 14 Working On or Near Live Conductors

This regulation states the requirement for working on or near equipment which has not been isolated and proved dead.

A typical example of live work would be live testing, e.g. the use of a suitable voltage indicator or multimeter on mains power.

The factors that have to be considered in deciding if it was justifiable for live work to proceed would include:

- When it is not practical to carry out the work with the conductors dead e.g. in the case of fault finding.

When working on or near conductors which are live suitable precautions would be:

- The use of people who are properly trained and competent to work on live equipment safely.

- The provision of adequate information about the live conductors involved.

- The use of suitable tools including insulated tools.

- The use of suitable insulated barriers or screens.

- The use of suitable test instruments and test probes.

Testing to establish whether electrical conductors are live or dead should always be done on the assumption they are live until such times as they have been proved dead.

When using test instruments or voltage indicators for this purpose they should be proved immediately before and after testing the conductors. It must be remembered that although live testing may be justified it does not follow that such justification can be made for the repair work to be carried out live. It should be carried out with the conductors proved dead.

Regulation 15 Working Space Access and Lighting

The purpose of this regulation is to ensure that sufficient space, access and adequate illumination is provided while persons are working on, at or near electrical equipment in order that they may work safely.

Regulation 16 Persons to be Competent to Prevent Danger and Injury

The object of this regulation is to ensure that persons are not placed at risk due to a lack of skills on the part of themselves or others in dealing with electrical equipment and the work associated with it.

The requirements are that persons must possess sufficient technical knowledge or experience or be supervised.

For the purpose of this regulation technical knowledge or experience may include the following:

- An adequate knowledge of electricity

- An adequate experience of electrical work

- An adequate understanding of the system to be worked on and practical experience of that system

- An understanding of the hazards which may arise and the precautions which need to be taken during work on a system

- An ability to recognise at all times if it is safe to continue to work

Regulation 29 Defence

This provides a defence for a duty holder who can establish that he has taken all reasonable steps and exercised all due diligence to avoid committing an offence.

Regulation 30 Exemption Certificates

This regulation allows the Health and Safety Executive to grant exemption certificates in special cases. This is unlikely to be required for individuals or organisations involved in the electrical systems for plumbing, heating and ventilating, refrigeration and gas installations.

Regulation 31 Extension Outside Great Britain

This concerns the application of the regulations outside Great Britain.

Regulation 32 Disapplication of Duties

This states the regulations do not apply to sea-going ships, hovercraft and aircraft but may apply to electrical equipment on vehicles if the equipment may give rise to danger.

Regulation 33 Revocations and Modifications

This regulation states the previous regulations which have been replaced or modified. (Includes Electricity (Factories Acts 1908 and 1944))

Appendix 1
Lists HSE and HSC Publications on Electrical Safety

Appendix 2
Lists other publications having an Electrical Safety Content

Appendix 3
Gives information concerning working space and access together with historical comment on revoked legislation.

BS 7671: 1992 Requirements for Electrical Installations (IEE Wiring Regulations 16th Edition)

The IEE Wiring Regulations have the status of a British Standard. The full title is British Standard 7671: 1992 Requirements for Electrical Installation (The IEE Wiring Regulations). The following statutory regulations recognise the IEE Wiring Regulations as a code of good practice.

Health and Safety at Work Act 1974
Electricity Supply Regulations 1988
Electricity at Work Regulations 1989

Plan and Style of Regulations

The 16th Edition is based on the international regulations produced by the International Electrotechnical Commission (I.E.C.). It is the aim of the I.E.C. to eventually have a common set of wiring regulations.

A large number of the new Regulations are based on I.E.C. rules already published. For some topics the I.E.C. work is still in progress, so the Regulations contain regulations from the previous edition. The 16th Edition requirements recognises British Standards, harmonised standards and foreign national standards based on an I.E.C. standard.

Format

The Regulations are divided into 7 parts with 6 Appendices. This format is illustrated overleaf.

Numbering

Each part of the Regulations is numbered consecutively, being identified by the first number of each group of digits. The parts are divided into chapters, identified by the second digit and each chapter is split into sections, identified by the third digit. After the first group of three digits, the digits separated by hyphens identify the regulation itself.

Example 1

| PART | CHAPTER | SECTION | SUB SECTION | REGULATION |

| 4 | 1 | 3 | - | 02 | - | 06 |

Example 2

Regulation number | 4 | 1 | 3 | - | 02 | - | 06 |

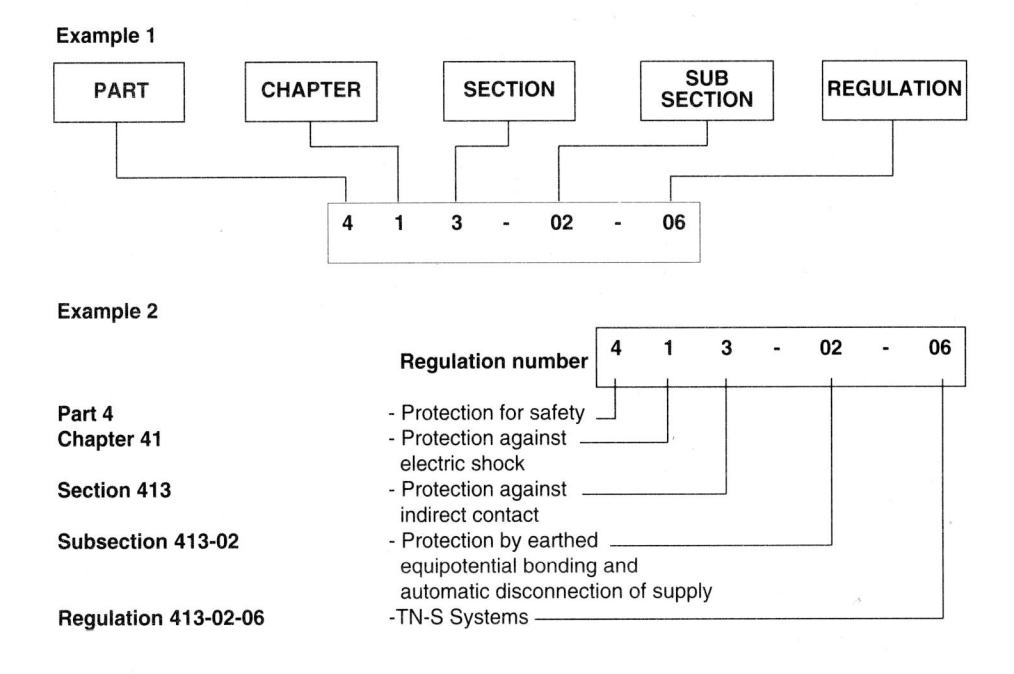

Part 4 — Protection for safety
Chapter 41 — Protection against electric shock
Section 413 — Protection against indirect contact
Subsection 413-02 — Protection by earthed equipotential bonding and automatic disconnection of supply
Regulation 413-02-06 — TN-S Systems

Index

At the beginning of each part of the Regulations, Chapters and Sections are listed; before each Chapter the Section and their Regulations are noted.

In their new form, the Regulations offer the designer the opportunity of complying with the Regulations by selecting, from various methods described, the one most suited to each particular installation from both economic and technical points of view.

As a result, the detail applying to a particular set of circumstances is sometimes spread across a number of widely separated regulations. None of these individual regulations can be applied in isolation; the overall situation can only be determined by taking into consideration all of the applicable regulations.

Cross-referencing related sections may be simplified:

- through the use of the index
   and
- by reference to the diagram on the following page

Plan of 16th Edition

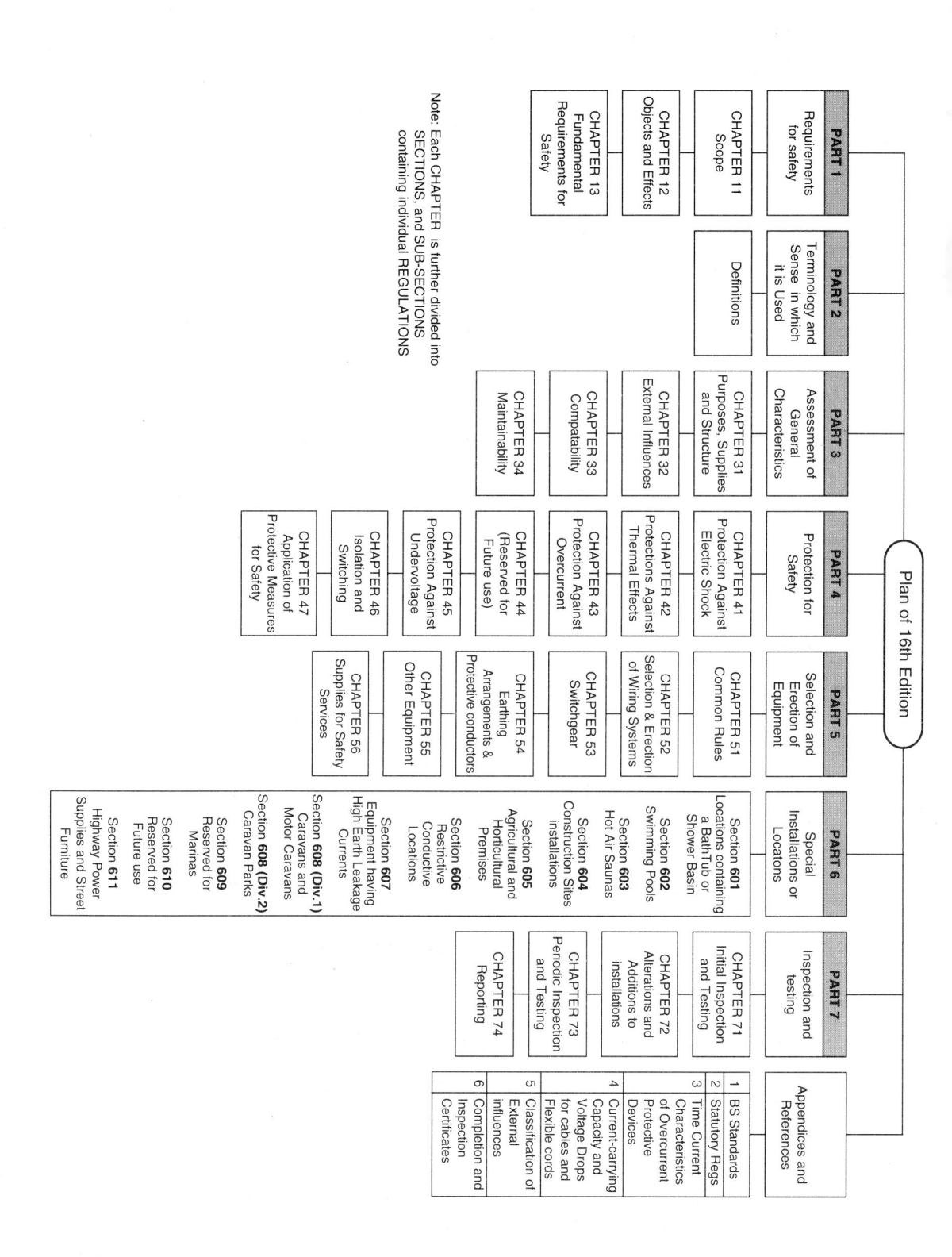

PART 1
- CHAPTER 11 Scope
- CHAPTER 12 Objects and Effects
- CHAPTER 13 Fundamental Requirements for Safety

Requirements for safety

PART 2 Terminology and Sense in which it is Used
- Definitions

PART 3 Assessment of General Characteristics
- CHAPTER 31 Purposes, Supplies and Structure
- CHAPTER 32 External Influences
- CHAPTER 33 Compatibility
- CHAPTER 34 Maintainability

PART 4 Protection for Safety
- CHAPTER 41 Protection Against Electric Shock
- CHAPTER 42 Protections Against Thermal Effects
- CHAPTER 43 Protection Against Overcurrent
- CHAPTER 44 (Reserved for Future use)
- CHAPTER 45 Protection Against Undervoltage
- CHAPTER 46 Isolation and Switching
- CHAPTER 47 Application of Protective Measures for Safety

PART 5 Selection and Erection of Equipment
- CHAPTER 51 Common Rules
- CHAPTER 52 Selection & Erection of Wiring Systems
- CHAPTER 53 Switchgear
- CHAPTER 54 Earthing Arrangements & Protective conductors
- CHAPTER 55 Other Equipment
- CHAPTER 56 Supplies for Safety Services

PART 6 Special Installations or Locations
- Section 601 Locations containing a BathTub or Shower Basin
- Section 602 Swimming Pools
- Section 603 Hot Air Saunas
- Section 604 Construction Sites installations
- Section 605 Agricultural and Horticultural Premises
- Section 606 Restrictive Conductive Locations
- Section 607 Equipment having High Earth Leakage Currents
- Section 608 (Div.1) Caravans and Motor Caravans
- Section 608 (Div.2) Caravan Parks
- Section 609 Reserved for Marinas
- Section 610 Reserved for Future use
- Section 611 Highway Power Supplies and Street Furniture

PART 7 Inspection and testing
- CHAPTER 71 Initial Inspection and Testing
- CHAPTER 72 Alterations and Additions to installations
- CHAPTER 73 Periodic Inspection and Testing
- CHAPTER 74 Reporting

Appendices and References
1. BS Standards
2. Statutory Regs
3. Time Current Characteristics of Overcurrent Protective Devices
4. Current-carrying Capacity and Voltage Drops for cables and Flexible cords
5. Classification of External Influences
6. Completion and Inspection Certificates

Note: Each CHAPTER is further divided into SECTIONS, and SUB-SECTIONS containing individual REGULATIONS

Scope, Object and Fundamental Requirements for Safety (Part 1)

Scope (Ref. Chapter 11)

The regulations apply to the:

- design
- selection
- erection
- inspection and testing of

electrical installations and include particular requirements for electrical installation of:

- locations containing a bath or a shower unit
- swimming pools
- locations containing a hot air sauna
- construction sites
- agricultural and horticultural premises
- restrictive conductive locations
- caravans and caravan parks
- highway power supplies and street furniture

In certain installations the requirements of the IEE Wiring Regulations may need to be supplemented by the requirements of the person ordering the work or by those of British Standards, e.g.

- emergency lighting - BS 5266
- installations in explosive atmospheres
- fire detection and alarm systems in buildings - BS 5839

Objects and Effects (Ref. Chapter 12)

The regulations are designed to protect persons, property and livestock from electric shock, fire and burns, also injury from mechanical movements of electrically-operated equipment.

They should be cited in their entirety if referred to in any contract.

They are not intended to take the place of detailed specifications, instruct untrained persons, or provide for every circumstance.

The advice of suitably qualified electrical engineers should be obtained for installations which are difficult or of a special character.

Compliance with Chapter 13 of the IEE Regulations will, in general, satisfy the statutory requirements listed in Appendix 2 of the regulations as follows:

Electricity Supply Regulations, 1988
Building Standards (Scotland) Regulations 1990
Electricity at Work Regulations 1989

Where statutory control of licensing is exercised the requirements of that authority should also be taken into account, e.g. local authority or fire authority.

Fundamental Requirements for Safety (Ref. Chapter 13)

These fundamental requirements for safety are identified under the following sub-headings, in very general terms as 20 short regulations. These are later greatly expanded upon in the body of Regulations and Appendices.

Workmanship and Proper Materials (130-01)

Good workmanship and proper materials shall be used.

General (130-02)

All equipment to be constructed, installed and maintained, inspected and tested for safety.

All equipment to be suitable for the maximum power demanded.

All conductors to be of sufficient size and capacity for the purposes intended.

All conductors to be either insulated, effectively protected, or placed and safeguarded to prevent damage as far as is reasonably practical.

All joints and connections to be properly constructed as regards conductance, insulation, mechanical strength and protection.

Overcurrent Protective Devices (130-03)

Installations and circuits must be protected against overcurrent by devices which will:

- operate automatically

- be of adequate breaking (and where appropriate) making capacity

- be suitably located and constructed to prevent danger from overheating, arcing, etc., and permit ready restoration of supply without danger.

Precaution Against Leakage and Earth Fault Currents (130-04)

Precautions are required where metalwork of electrical equipment may become charged with electricity, due to a breakdown in insulation or a fault in equipment, in such a way as to cause danger:

- metalwork is to be earthed in order to discharge electrical energy without danger or protected by other, equally effective, means.

Every circuit to be arranged to prevent persistent dangerous earth leakage currents.

Where metalwork of electrical equipment is earthed, the circuits concerned are to be protected against dangerous earth fault currents by overcurrent protective devices such as fuses and miniature circuit breakers or residual current devices, or any other equally effective device.

Where metalwork of electrical equipment is earthed and is accessible at the same time as other exposed metal parts of other services, (e.g. gas and water), these other services must be effectively connected to the main earthing terminal of the installation, if the metalwork of the other services is liable to introduce a potential, generally earth potential.

Protective Devices and Switches (130-05)

No fuse or circuit breaker, other than a linked circuit breaker, shall be inserted in an earthed neutral conductor.

Single-pole switches shall be inserted in the phase conductor only, and any switch so arranged that it also breaks all the related phase conductors.

Isolation and Switching (130-06)

Isolation and switching arrangements shall be installed so that all voltage can be effectively cut-off from every installation or from every circuit and from all equipment as may be necessary to prevent or remove danger. These must be effective and positioned for ease of operation.

Every electric motor shall be provided with an efficient means of disconnection, which shall be readily accessible, easily operated and placed so as to prevent danger.

Accessibility of Equipment (130-07)

Every piece of equipment which requires operation or attention shall be installed so that an adequate and safe means of access and working space is afforded for such operation or attention.

Precautions in Adverse Conditions (130-08)

All equipment likely to be exposed to corrosive atmospheres and adverse weather or other conditions shall be constructed or protected to prevent danger arising from such exposure.
Where there is risk of fire or explosion all equipment shall be constructed or protected, and other special precautions taken, to prevent danger.

Additions and Alterations to an Installation (130-09)

No additions or alterations shall be made to an existing installation unless it has been ascertained that the ratings and condition of any existing equipment (including the supply) is adequate for the altered circumstances and any additional load, and that the earthing arrangements are also adequate.

Inspection and Testing (130-10)

On completion of an installation, or extension or alteration of any installation, appropriate tests and inspection should be made to verify, so far as is reasonably practicable, that the requirements of Regulations 130-01 to 09 have been met.

Persons ordering or requesting an inspection should be informed of the requirements for periodic inspection and testing covered in Chapter 13 of the regulations by the person carrying out the inspection.

Assessment of General Characteristics (Part 3)

General

This part of the regulations deals with the need to assess the general characteristics of the energy source or supply and the installation itself.

The following factors are to be assessed and taken into account when choosing the methods of protection for safety (Part 4) and when selecting and erecting equipment (Part 5).

- the purpose for which the electrical installation is to be used, its general structure and supplies (Chapter 31)
- the external influences to which it is exposed (Chapter 32 and Appendix 5)
- the compatibility of its equipment (Chapter 33)
- its maintainability (Chapter 34)

Purposes, Supplies and Structure (Ref. Chapter 31)

Maximum Demand and Diversity

The maximum demand of the electrical installation expressed as a current value, must be assessed. Diversity may be taken into account when determining the maximum demand.

Arrangement of Live Conductors and Types of Earthing

The characteristics for number and type of live conductors and earthing arrangements must be assessed and the appropriate methods of protection for safety selected to avoid danger.

Number and Type of Live Conductors e.g. Single-phase 2 wire or three-phase 4 wire (a.c.)

The number and types of live conductors, e.g. single-phase 2 wire or three-phase 4 wire (a.c.) for the source of energy and for the circuits to be used in the installation need to be assessed; the energy supplier (e.g. electricity company) should be consulted where necessary.

Type of Earthing Arrangement

The type of earthing arrangement(s) to be used must also be determined. The choice of arrangements may be limited by the characteristics of the energy source and any facilities for earthing.

Nature of Supply

The following characteristics should be ascertained for an external supply (and be determined for a private source)

(i) nominal voltage(s)

(ii) the nature of current and frequency

(iii) the prospective short-circuit current at the origin of the installation

(iv) the earth loop impedance of that part of the system external to the installation

(v) suitability for the requirement of the installation, including maximum demand

(vi) type and rating of the overcurrent protective device at the origin of the installation

Installation Circuit Arrangements

Every installation should be divided into circuits as necessary to:

- avoid danger in the event of a fault, and

- facilitate safe operation, inspection, testing and maintenance

A separate circuit shall be provided for each part of an installation which has to be separately controlled to prevent danger, so that the circuits remain energised in the event of failure of other circuits in the installation; e.g. emergency stop circuit controlling the power supply in a workshop, when operated, cuts off the power to the machines but not the lighting which has to be separately maintained to prevent danger.

The number of final circuits required in an installation and the number of points supplied by a final circuit shall be arranged to comply with the requirements for overcurrent protection (Chapter 43), isolation and switching (Chapter 46), and current-carrying capacity of conductors (Chapter 52).

Each final circuit must be connected to a separate way in a distribution board, and the wiring of each final circuit should be electrically separate from every other final circuit.

External Influences (Ref. Chapter 32)

The chapter dealing with the external influences likely to effect the design and safe operation of the installation is not yet at a stage for adoption as a basis for national regulations. Appendix 5 of the regulations contains some useful information on the subject.

Compatibility (Ref. Chapter 33)

An assessment should be made of any characteristics of equipment likely to have harmful effects upon other electrical equipment or other services or likely to impair the supply.

The following characteristics have (for example) been identified:

- transient overvoltage
- rapidly fluctuating loads
- starting currents
- harmonic currents (e.g. fluorescent lighting loads and thyristor drives)
- earth leakage currents

Maintainability (Ref. Chapter 34)

Maintainability is also a very important factor to consider when deciding on the design on an installation.

An assessment shall be made of the frequency and quality of maintenance that the installation can reasonably be expected to receive during its intended life. This shall include (where practicable) consultation with the person or body responsible for the operation and maintenance of the installation. Only then can the regulations be applied so that:

- any periodic inspection, testing, maintenance and repairs likely to be necessary during the intended life can be readily and safely carried out and
- the protective measures for safety remain effective during the intended life and
- the reliability of equipment is appropriate to the intended life.

Electrical Isolation

Introduction

Electricity, safely controlled, is a very efficient and convenient way of distributing and using energy. If it is inadequately controlled it can be lethal.

Electric Shock

Is an effect on the nervous system with a resulting contraction of muscles and feeling of concussion.

An electric shock can be received by either direct or indirect contact.

Direct Contact - Contact with parts or conductors intended to be energised in normal use.

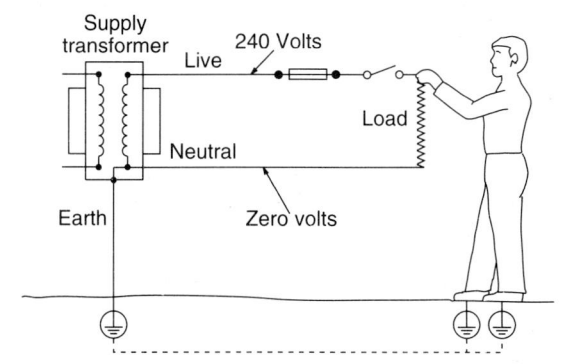

Indirect Contact - Contact with exposed or extraneous conductive parts that may be made live by a fault.

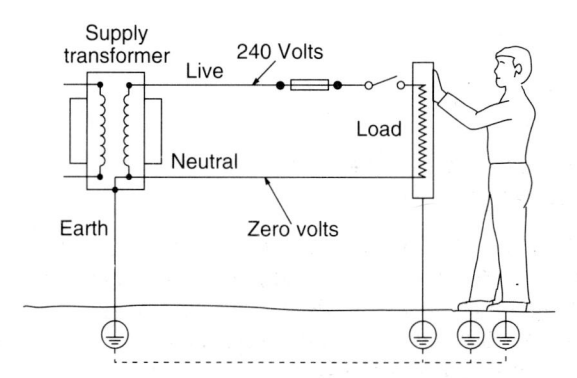

Body Resistance

This varies due to the conditions of the body, age, situation and weather.
Up to 50 volts a.c. is the maximum voltage level allowed for safety.
This value is referred in the BS 7671 (The IEE Wiring Regulations).

Perception Level

1mA is the point at which an individual is aware of electric current.

Let Go Level

9mA is the point at which an individual still has control over the effects of shock on the muscles in the body to be able to release a gripped conductor.

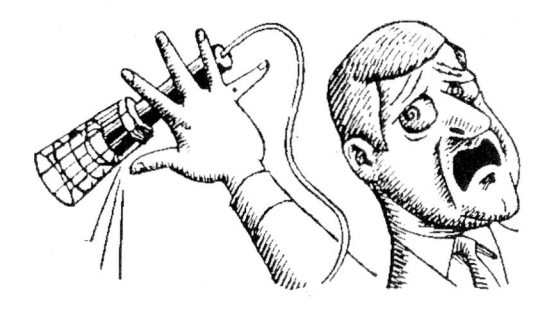

Freezing and Muscular Contraction

With further increases in levels of current (around 20mA) the subject cannot release themselves. Extreme pain is felt which may cause the subject to lose consciousness, or the body muscles may contract affecting the lungs and the subject may die from asphyxia.

Death

With currents of around 80mA, death is likely to occur by ventricular fibrillation and severe internal and external burns.

Ventricular Fibrillation

This is the effect of electric shock which causes the muscles of the heart to contract separately at different times instead of in unison. This condition is a killer.

Electrocardiogram showing
Normal Heart Beat

Electrocardiogram showing
Ventricular Fibrillation

Removing Persons

Removing a person from contact with live conductors needs great care. The following points should be considered:

- The rescuer must not be put in danger
- Procedures must be undertaken as quickly as possible
- Procedures to be carried out in a way that prevents further injury

Disconnect the electricity supply first wherever possible. Pull victim away from live conductors using dry clothing such as overalls wrapped around them to enable them to be effectively and quickly removed.

Treatment

Summon assistance and call for an ambulance. In the case of slight shock reassure patient, make them comfortable and report accident to appropriate personnel.

If burns have been sustained, cool the areas with cold water or any other non-flammable fluid at hand. Remove anything of a constrictive nature if possible, such as rings, belts and boots. If burns are serious, cool the areas and send to hospital without delay.

For severe cases of shock with victim unconscious and not breathing, clear the airway and administer mouth to mouth resuscitation - remember there is no time to waste as lack of oxygen to the brain can cause damage within 4 minutes.

On restoration of breathing place the patient in the recovery position and send to hospital without delay.

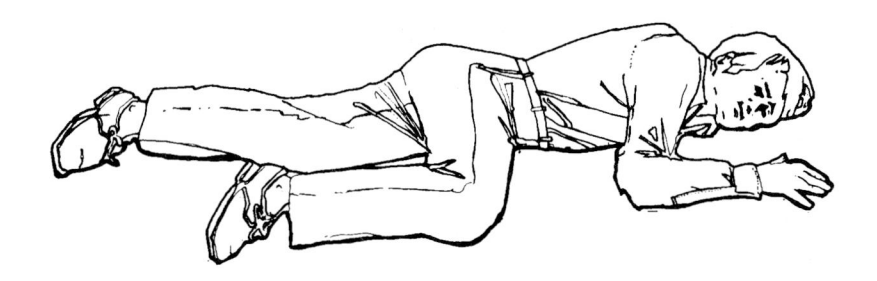

In the case of the heart being stopped then cardiac compression should be given.

Electrical Isolation

Before beginning work on any electrical circuit you should make sure that it is completely isolated from the supply. Electrically powered machines are usually fitted with an isolator for disconnecting the supply under 'no-load' conditions. Otherwise it may be necessary to isolate the supply by removing fuses, locking-off MCBs or by physical disconnection of live conductors.

Fuses, switch-fuses and isolators should be clearly marked to indicate the circuit they protect.

Any isolating device, when operated, should be capable of being locked in the open position or carry a label stating otherwise. If the isolator consists of fuses these should be removed to a safe place where they cannot be replaced without the knowledge of the responsible person concerned. For example, they can be kept in the pocket if the job is of short duration, or in a locked cupboard provided for the purpose in the charge of the works or site supervisor.

Fuses should never be removed or replaced without first switching off the supply.

Your Responsibility

When working on a particular circuit or appliance you must be certain that the supply cannot be switched on without your knowledge. You must satisfy yourself that the circuit is open and labelled so.

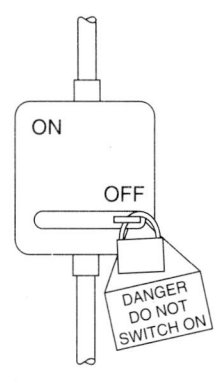

Test Lamps

Neon testers should not be relied on since the neon will not indicate supplies at low potentials. There is also the risk of receiving an electric shock if the resistor breaks down.

Flex and lampholders (homemade) test lamps should never be used. These are extremely dangerous, since mains potential is present in all the components; flex and bulb are vulnerable to damage, and there is usually no fuse in circuit.

Approved test lamps have a robust plastic body containing the circuitry. Resistors are fitted in the circuit to limit the potential and fuses are fitted to give complete protection in case of any fault occurring. The exposed metal probe is kept to a minimum distance i.e. 1.00mm.

Always check that the circuit is dead, using approved test lamps, to prevent danger from electric shock.

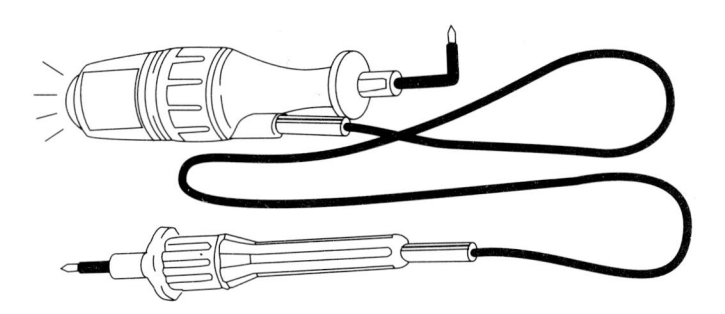

Remember always prove your voltage indicating device on a known supply or voltage proving unit immediately before and immediately after testing the conductor or equipment for safe and effective isolation.

Voltage Proving Unit

A unit providing a 240 volt output from a 9 volt d.c. battery input to prove the integrity of neon-type test lamps.

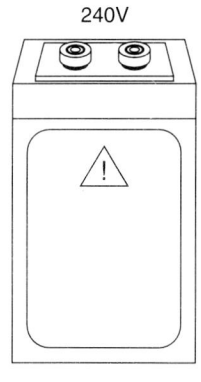

Local Isolation

It has been the practice of British Gas to ask for the electrical supply point, which will be used to supply a central heating system to be installed as close as practical to the boiler and in a readily accessible position. The type of supply point being a 13 amp BS 1363 socket outlet and plug top. The combination provides local isolation.

A more acceptable method of providing local isolation would be to install a double-pole switch fuse spur unit, in which the fuse carrier can be withdrawn but not removed and a small padlock fitted as illustrated.

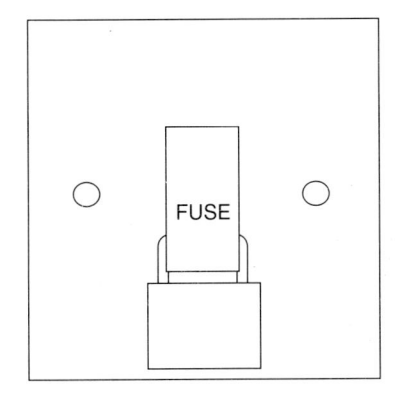

Isolation and Switching

The IEE Regulations state that a means of isolation is required for circuits and equipment in order to enable a skilled person to carry out work on that installation or equipment safely with it dead.

All isolating devices must comply with British Standards which means the position of the contacts or other means of isolation must be externally visible or clearly indicated.

For TN-S and TN-C-S systems all phase conductors must be switched and in TT systems all live conductors.

The main switch in a consumer unit which is a double-pole switch is an isolating device which is capable of isolating all the circuits.

Functional Switching

The IEE Regulations also require a means of interrupting the supply on load for every circuit, example a switch for a lighting point or immersion heater. Remember that in bathrooms insulated cord-operated switches should be used.

General

All isolating and switching devices should be located in readily accessible positions to avoid danger to the operator.

ISOLATING A COMPLETE INSTALLATION

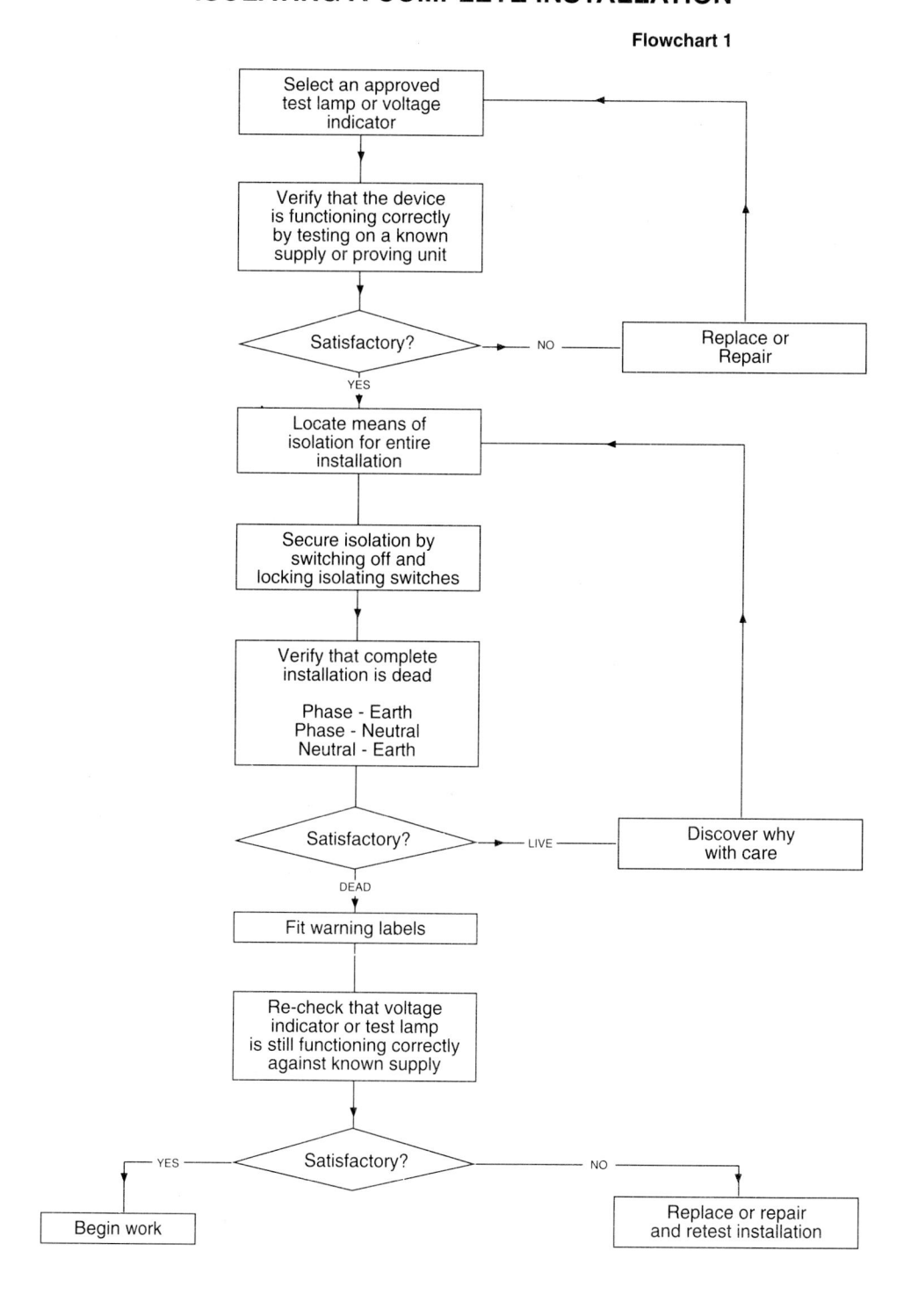

Flowchart 1

```
            ┌─────────────────────────┐
            │ Select an approved      │◄──────────────────┐
            │ test lamp or voltage    │                   │
            │ indicator               │                   │
            └───────────┬─────────────┘                   │
                        ▼                                 │
            ┌─────────────────────────┐                   │
            │ Verify that the device  │                   │
            │ is functioning correctly│                   │
            │ by testing on a known   │                   │
            │ supply or proving unit  │                   │
            └───────────┬─────────────┘                   │
                        ▼                                 │
                  ◄ Satisfactory? ►──── NO ──►  ┌──────────────────┐
                        │                       │ Replace or Repair│
                       YES                      └──────────────────┘
                        ▼
            ┌─────────────────────────┐
            │ Locate means of         │◄──────────────────┐
            │ isolation for entire    │                   │
            │ installation            │                   │
            └───────────┬─────────────┘                   │
                        ▼                                 │
            ┌─────────────────────────┐                   │
            │ Secure isolation by     │                   │
            │ switching off and       │                   │
            │ locking isolating switches                  │
            └───────────┬─────────────┘                   │
                        ▼                                 │
            ┌─────────────────────────┐                   │
            │ Verify that complete    │                   │
            │ installation is dead    │                   │
            │   Phase - Earth         │                   │
            │   Phase - Neutral       │                   │
            │   Neutral - Earth       │                   │
            └───────────┬─────────────┘                   │
                        ▼                                 │
                  ◄ Satisfactory? ►── LIVE ──►  ┌──────────────────┐
                        │                       │ Discover why     │
                      DEAD                      │ with care        │
                        ▼                       └──────────────────┘
            ┌─────────────────────────┐
            │ Fit warning labels      │
            └───────────┬─────────────┘
                        ▼
            ┌─────────────────────────┐
            │ Re-check that voltage   │
            │ indicator or test lamp  │
            │ is still functioning correctly │
            │ against known supply    │
            └───────────┬─────────────┘
                        ▼
         ┌──YES──◄ Satisfactory? ►── NO ──┐
         ▼                                ▼
   ┌──────────┐                 ┌──────────────────────┐
   │Begin work│                 │ Replace or repair    │
   └──────────┘                 │ and retest installation│
                                └──────────────────────┘
```

ISOLATING AN INDIVIDUAL CIRCUIT OR ITEM OF FIXED EQUIPMENT

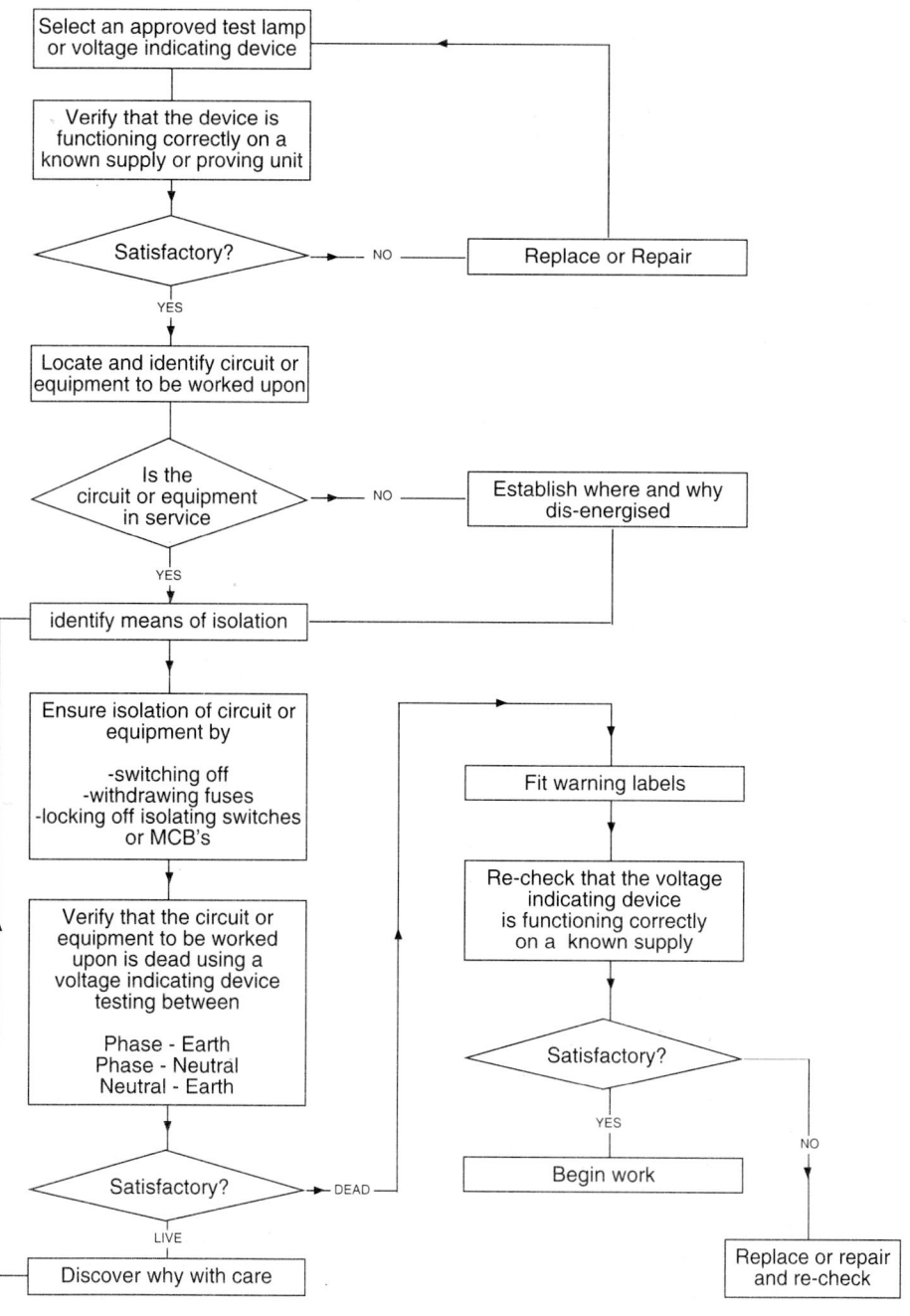

Flowchart 2

```
Select an approved test lamp or voltage indicating device
        │
        ▼
Verify that the device is functioning correctly on a known supply or proving unit
        │
        ▼
    ◇ Satisfactory? ──NO──▶ Replace or Repair ──▶ (back to Select)
        │ YES
        ▼
Locate and identify circuit or equipment to be worked upon
        │
        ▼
    ◇ Is the circuit or equipment in service ──NO──▶ Establish where and why dis-energised
        │ YES
        ▼
identify means of isolation
        │
        ▼
Ensure isolation of circuit or equipment by
  -switching off
  -withdrawing fuses
  -locking off isolating switches or MCB's
        │
        ▼
Verify that the circuit or equipment to be worked upon is dead using a voltage indicating device testing between
  Phase - Earth
  Phase - Neutral
  Neutral - Earth
        │
        ▼
    ◇ Satisfactory? ──DEAD──▶ Fit warning labels
        │ LIVE                    │
        ▼                         ▼
Discover why with care    Re-check that the voltage indicating device is functioning correctly on a known supply
                                  │
                                  ▼
                              ◇ Satisfactory?
                               │ YES      │ NO
                               ▼          ▼
                           Begin work   Replace or repair and re-check
```

Earthing Arrangements and Protective Conductors

Earthing Arrangements

General

The earth can be considered to be a large conductor which is at zero potential. The purpose of earthing is to connect together all metal work (other than that which is intended to carry current) to earth so that dangerous potential differences cannot exist either between different metal parts, or between metal parts and earth, for example, in a kitchen the electric refrigerator and a central heating radiator or the electric washing machine and the water pipes.

Purpose of Earthing

By connecting to earth metalwork not intended to carry current, a path is provided for leakage current which can be detected and, if necessary, interrupted by the following devices:

- fuses
- circuit breakers
- residual current devices

Where a building in which an electrical installation is being carried out is protected by a lightning protection system, account must be taken of the requirements of BS 6651 (Code of Practice for Protection of Structures Against Lightning).

Connections to Earth

The earthing arrangement of an installation must be such that:

- The value of impedance from the consumer's main earthing terminal to the earthed point of the supply for TN systems or to earth for TT systems is in accordance with the protective and functional requirements of the installation and expected to remain continuously effective.
- earth fault and earth leakage currents which may occur under fault conditions can be carried without danger, particularly from any heat and magnetic forces created.
- They are robust and protected from mechanical damage appropriate to the assessed conditions.

The installation should be so installed as to avoid risk of subsequent damage to any metal parts or structures through electrolysis.

Main Earthing Terminal

A main earthing terminal or bar must be provided in an accessible position for every installation, for the connection of the circuit protective conductors, main bonding conductors, functional earthing conductors and any lightning protection system bonding conductors. Provision must also be made for disconnection of the main earthing terminal from the means of earthing to permit measurement of the resistance of the earthing arrangements.

The method of disconnecting the earthing terminal from the means of earthing must be such that it can only be effected with the use of tools for example, a screwdriver.

Earthed Equipotential Bonding

This system of protection is one most commonly used.

The System

The aim is that all exposed metalwork of the electrical installation, the appliances connected to it, and metalwork in the building which is in contact with earth such as pipes, taps, baths and sinks are joined together by protective conductors to a main earthing terminal. This terminal is usually provided by the electricity company but in some cases where the supply is by overhead wires the earth may be provided by means of a copper rod driven into the ground. If, for example, the metal casing of an electrical appliance becomes live because of a loose wire inside it, the current will flow harmlessly to earth through the protective conductor which is connected to it. If the metal was not earthed and you touched it, the current would instead pass through your body to earth, giving you a potentially lethal electric shock.

Typical Arrangement Illustrating Protective Conductor

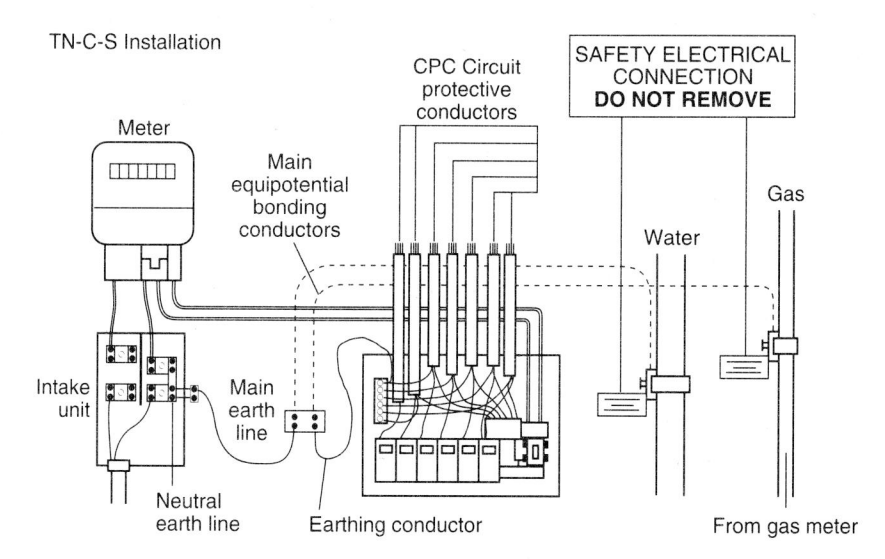

Earthing Conductors

For TN-C and TN-C-S systems the size of earthing conductor is determined by the local electricity company and is usually installed by them.

Main Equipotential Bonding Conductors

The purpose of installing bonding conductors is to ensure that any metalwork within an installation, such as gas and water services, are at the same potential as the metalwork of the electrical installation.

This is achieved by installing main bonding conductors from the main earthing terminal to the gas, water and other services at the point of entry to the premises as illustrated below.

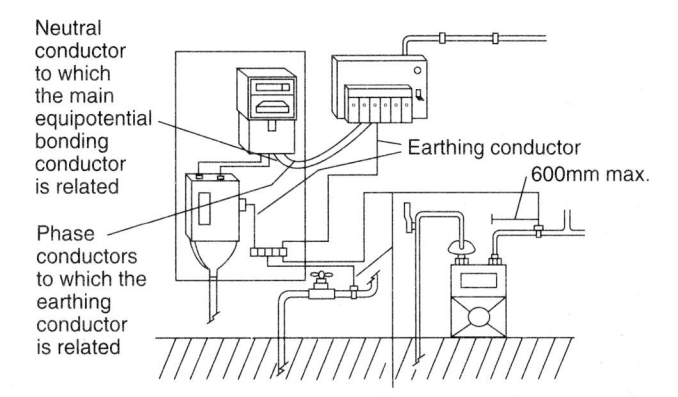

Bonding of gas service pipes should be made on the consumer's side of the meter between the meter outlet union and any branch pipework, but within 600mm of the gas meter as illustrated.

If there is a water meter it should be shunted by a bonding conductor, to prevent damage to the working parts of the meter in the event of a fault current flowing through that section of pipework in which the meter is connected.

Note: Check with local electricity, gas and water companies for any special requirements regarding the bonding of services. Also see Table 54H IEE Regulations.

Typical Values

Cross Sectional Area of Supply Neutral Copper Conductor	Minimum Cross Sectional Area of Main Equipotential Bonding Conductor
$35mm^2$ or less	$10mm^2$
Over $35mm^2$ up to $50mm^2$	$16mm^2$
Over $50mm^2$ up to $90mm^2$	$25mm^2$

Typical Earthing and Bonding Conductor Arrangements

TN-S earthed to armour or metal sheath or electricity company's cable.

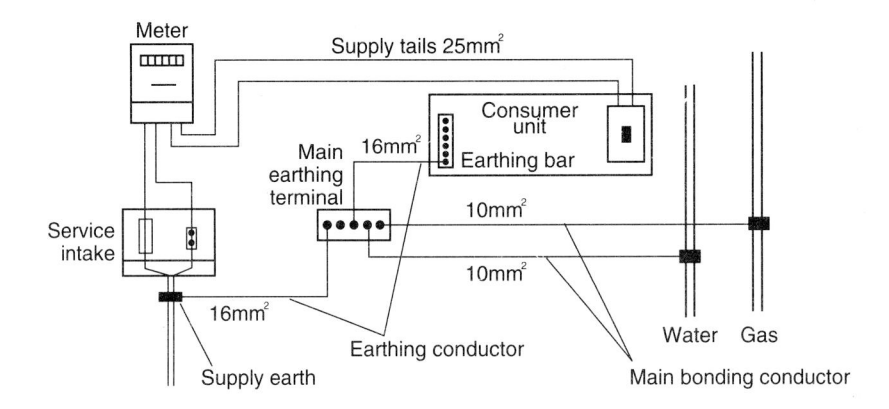

TN-C-S earthed using combined neutral and earth conductor of electricity company's cable

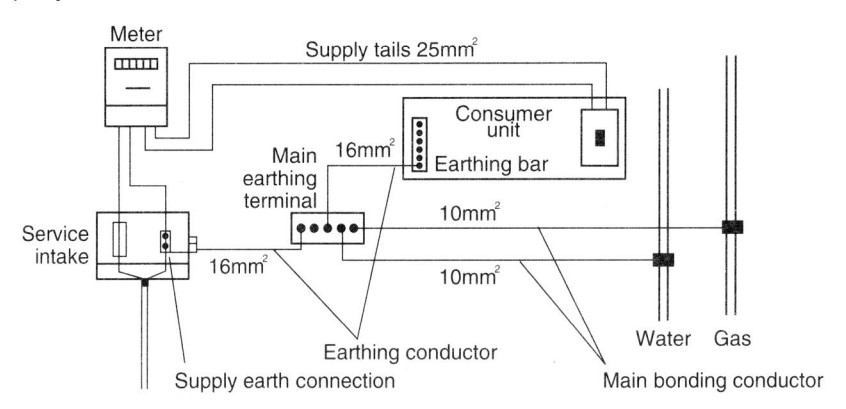

TT earthed via an earth electrode

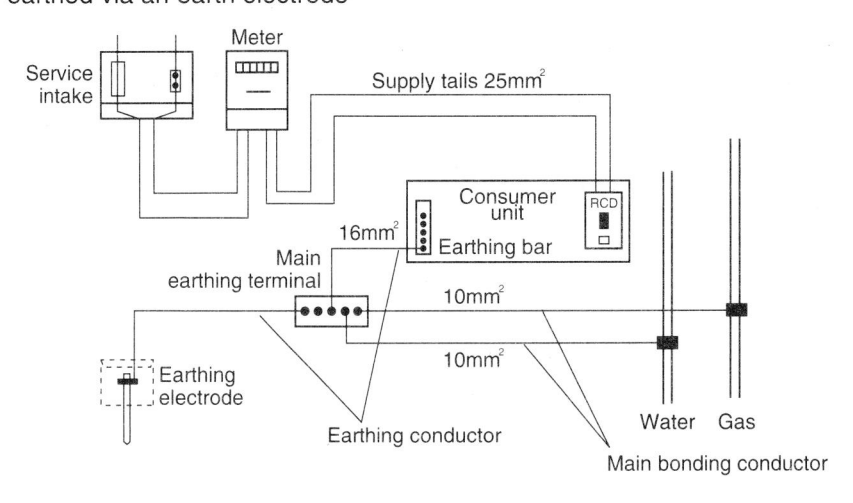

Main bonding of industrial/commercial installations

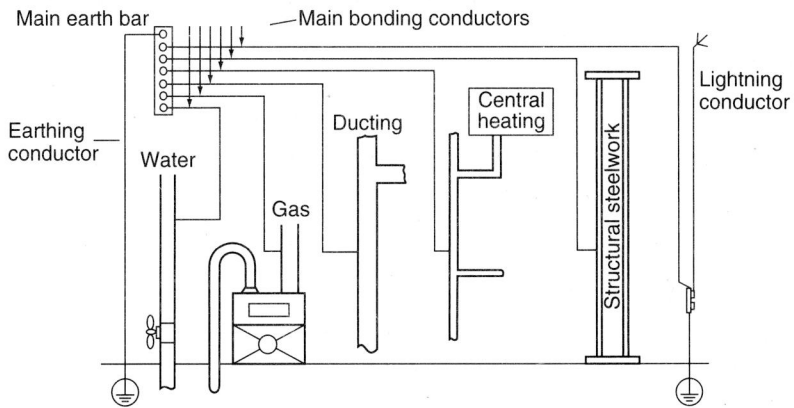

Types of Circuit Protective Conductors

- PVC insulated single core cable manufactured to BS 6004 colour green/yellow
- PVC insulated and sheathed cable with an integral protective conductor manufactured to BS 6004
- Copper strip
- Metal conduit
- Metal trunking systems
- Metal ducting   enclosures
- MICC - cable sheath
- Lead covered cable sheath
- SWA cable armourings

When the protective conductor is formed by a wiring system such as conduit, trunking, MICC, armoured cables or sheathed and insulated cables, a separate protective conductor must be installed from the earthing terminal of socket outlets to the earthing terminal of the associated box as illustrated.

The circuit protective conductor of final ring circuits which are not formed by the metal covering or enclosures of a cable should be installed in the form of a ring having both ends connected to the earth terminal at the origin of the circuit, e.g. distribution board or consumer's unit as illustrated.

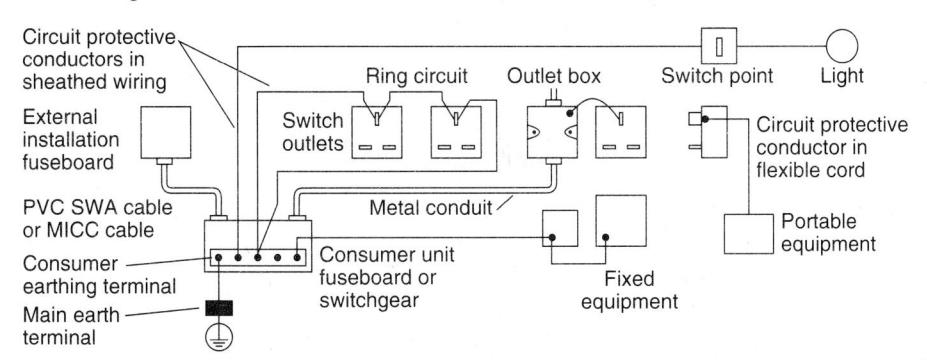

Exposed conductive parts of equipment should not be used to form part of protective conductors for other equipment, except if the exposed conductive part is a metal enclosed busbar trunking system or similar enclosures of actory-built assembly, e.g. the case of a distribution fuse board, which may form part of a conduit system and hence be a part of the protective conductor circuit and which is constructed to satisfy the following requirements:

- the electrical continuity is achieved so as to afford protection against mechanical, chemical and electro-chemical deterioration.
- There is a provision for the connection of other protective conductors at tap-off points

Flexible Conduit

Flexible conduit cannot be used as a protective conductor; an additional circuit protective conductor should be installed within the conduit but be accessible at the terminations.

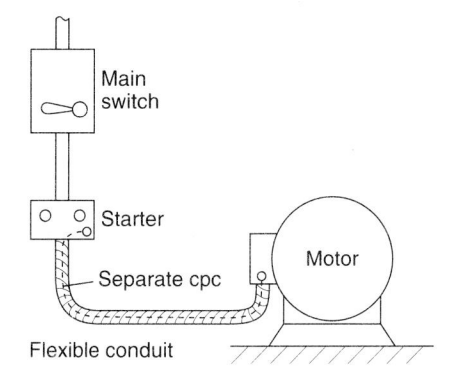

Cross Sectional Areas of Circuit Protective Conductors

The minimum cross sectional area of protective conductors can be obtained by using Table 54G or by use of the formula

$$S = \sqrt{\frac{I^2 t}{k}} \text{ mm}^2$$

where: S is the cross sectional area in mm^2
I value of maximum fault in amperes
t operating time of the device in seconds
k factor for specific protective conductor
(Tables 54B-F of the IEE Regulations)

If the protective conductor does not form part of a cable, is not a conduit, ducting or trunking, and is not contained in an enclosure formed by the wiring system, for example, a separate conductor, the cross-sectional area should not be less than:

- $2.5mm^2$ if sheathed, or otherwise provided with mechanical protection
- $4mm^2$ where mechanical protection is not provided

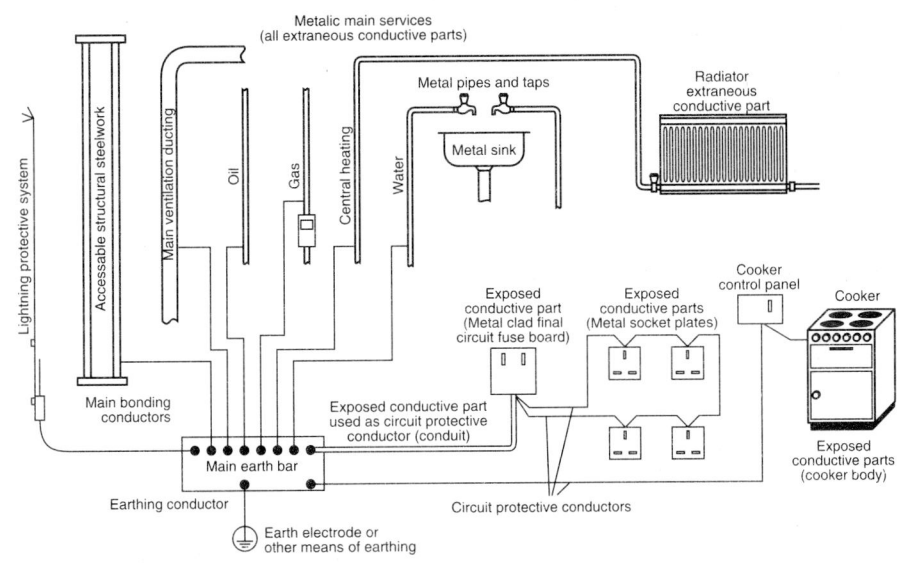

Preservation of Electrical Continuity of Protective Conductors

Protective conductors should be installed so that they are protected against mechanical damage, chemical deterioration and electrodynamic effects.

Where protective conductors of cross-sectional areas up to and including $6mm^2$ are installed, which are not an integral part of a cable or a cable enclosure such as conduit or trunking, they must be protected throughout by insulation at least equivalent to that provided for a single-core non-sheathed cable of appropriate size complying with BS 6004 or BS 7211.

When the sheath of a cable containing an insulated protective conductor is removed for the purpose of making terminations or joints, protective conductors up to and including $6mm^2$ must be protected by a green/yellow insulating sleeve complying with BS 2848.

Connections of protective conductors should be accessible for inspection. This does not apply to:

- a compound filled/encapsulated joint
- joints made by welding, soldering, brazing or compression tools
- joints made in metal conduit, ducting or trunking system

All joints in protective conductors must be electrically and mechanically sound. Joints in metal conduit systems should be screwed or mechanically clamped; plain slip or pin grip sockets are not suitable as they will not ensure an effective low resistance joint throughout the life of the installation due to the small area of contact achieved with this type of fitting.

No switch should be inserted in a protective conductor.

Supplementary Bonding Conductors

A supplementary bonding conductor used to connect exposed conductive parts must have a cross-sectional area not less than the smallest protective conductor connected to the exposed conductive parts, subject to a minimum of:

- 2.5mm^2 if sheathed or mechanically protected

- 4mm^2 if mechanical protection is not provided

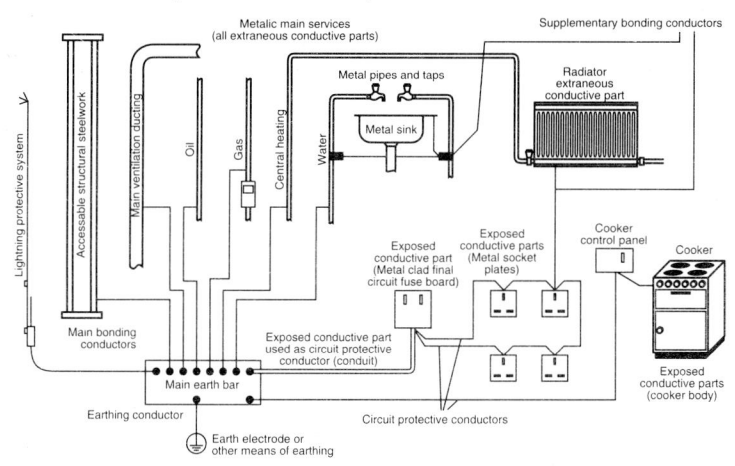

In situations where a supplementary bonding conductor connects two extraneous conductive parts, neither of which are connected to an exposed conductive part, the minimum cross-sectional area of the supplementary bonding conductor shall be:

- 2.5mm^2 if sheathed or mechanically protected

- 4mm^2 if mechanical protection is not provided

Supplementary bonding conductors may need to be installed in situations such as kitchens where a person may make simultaneous contact with an electrical appliance (such as an electric kettle or washing machine) and other metalwork, (such as the hot or cold water taps). In these situations supplementary bonding may be required as illustrated (but not if earth continuity tests prove that all metalwork of electrical, gas and water services is effectively bonded). The pipework of the gas and water services may be effectively connected together using permanent and reliable metal to metal joints of negligible impedance.

Supplementary bonding conductor

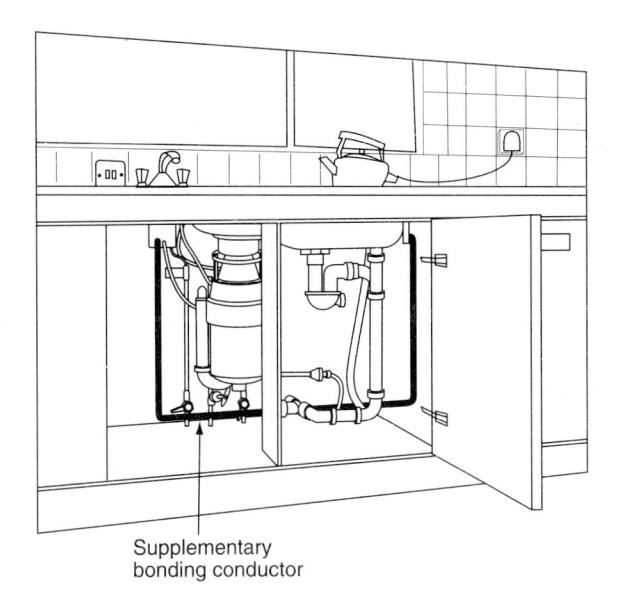

A generally acceptable test which can be used to establish if an item is an extraneous conductive part is as follows.

Measure the insulation resistance using a 500V d.c. insulation resistance test meter between the item and the main earthing terminal of the installation. If the value indicated on the test meter is greater than 0.25 Megohm and a visual inspection confirms that this value is not likely to deteriorate, the item can be as far as is reasonably practical considered not to be an extraneous conductive part.

This situation is based on the principle that 1mA flowing through the human body is not considered dangerous and with resistance values of $0.25M\Omega$ or greater on a 240 volt supply system this value of current will not be exceeded.

If the insulation resistance test indicates negligible resistance and this was confirmed by using an ohmmeter then the item may not need to be bonded.

If the resistance is between these values (i.e. $>1\Omega$ $<0.25M\Omega$) then the item should be bonded.

Rooms Containing Fixed Bath or Showers

In rooms containing a fixed bath or shower where permanent and reliable metal pipework has been installed, supplementary bonding conductors shall be installed as illustrated below, to reduce to a minimum the risk of electric shock in circumstances when the body resistance is likely to be low. The most common method of making a bonding connection to pipework is by using BS 951 earth clips.

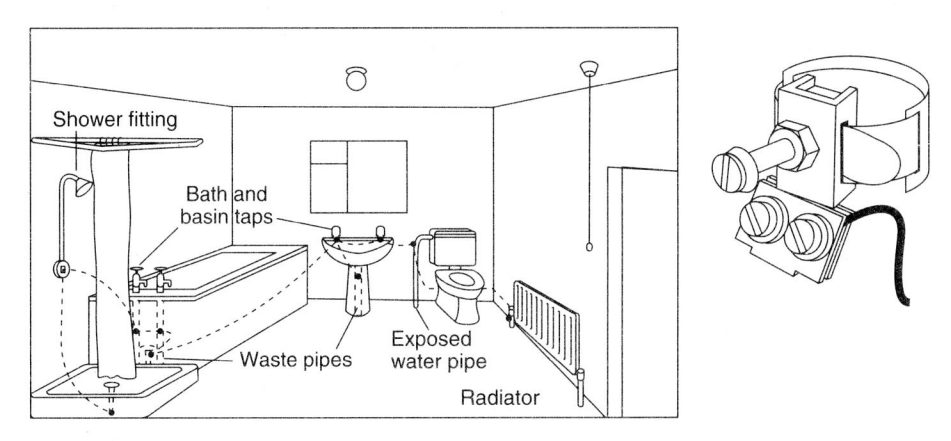

In other cases it is necessary to bond hot, cold and heating system pipework at one point only e.g. airing cupboard. It should be remembered that supplementary bonding should be neat and tidy and not cause injury, so the requirement that needs to be remembered is that connections should be accessible for inspection and not necessarily obtrusively visible.

Termination to Earth Electrodes

The connection of earthing conductors to electrodes require adequate insulation where they enter the ground, to avoid possible dangerous voltage gradient at the surface. All electrode connections must be thoroughly protected against corrosion and mechanical failure.

It is important that the electrode is made accessible for inspection purposes, and a label should be fitted at or near the point of connection.

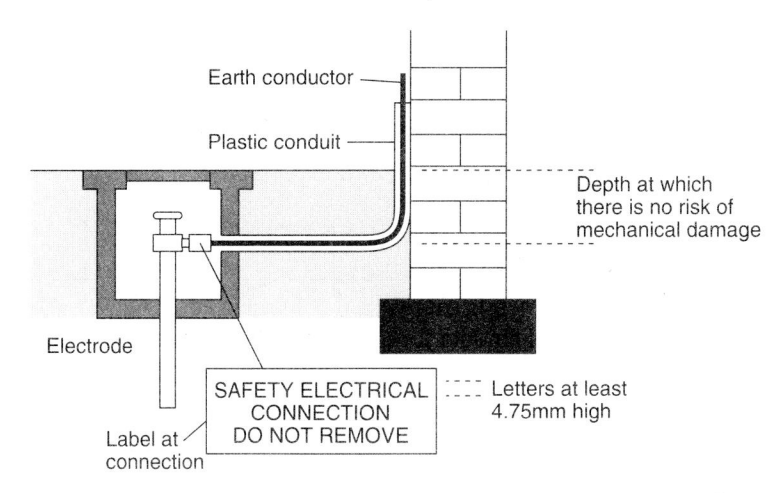

Installing an Earth Electrode

Earth Electrode Resistance Area

Every earth electrode has a definite electrical resistance to earth. Current flowing from the electrode to the general mass of earth has to traverse the concentric layers of soil immediately surrounding the electrode. Since the soil is a relatively poor conductor of electricity and as the cross-sectional areas of the layers of soil nearest to the electrode are small, the result is that of a graded resistance concentrated mainly in the area of soil surrounding the electrode. Moreover, the surface of the soil near the electrode will become "live" under fault conditions.

Surface Voltage Gradients

The illustration shows a typical surface voltage distribution near an earth electrode. The cow standing on the ground near the "live" electrode may receive a considerable voltage between its fore and hind feet resulting in a dangerous and possibly lethal shock since voltages of around 25V are dangerous to livestock.

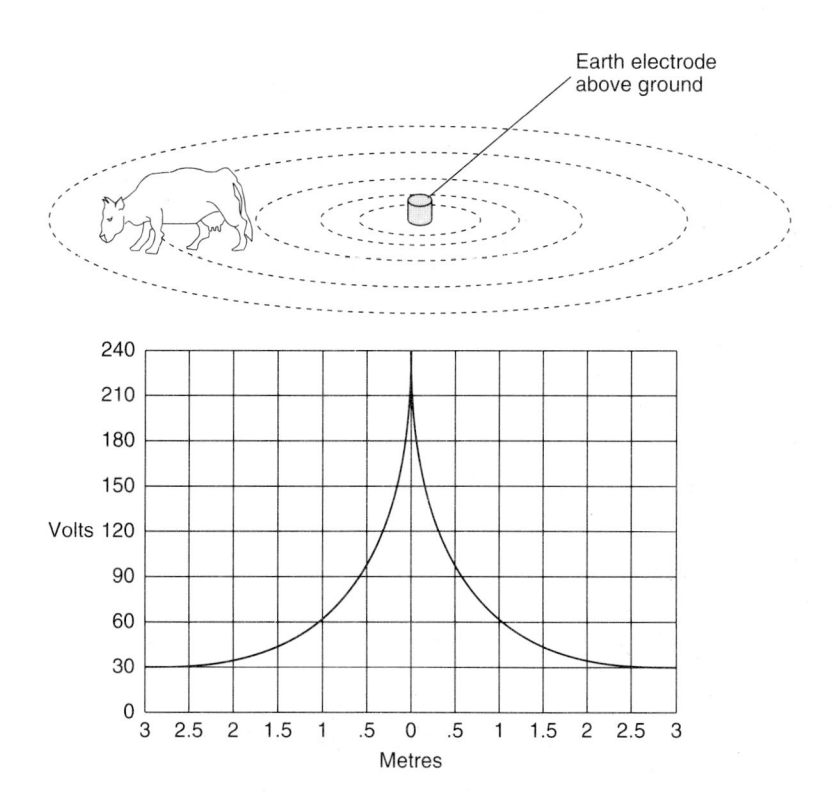

Electrical Cables and Accessories

Wiring Accessories

The range of different wiring accessories used for wiring circuits is wide. If we consider, for instance, lighting circuits, you will find ceiling roses connected by a flex to lampholders to form a pendant or batten lampholders for direct fixing to the ceiling and of course, switches for turning the lights on and off. These may be wall or ceiling mounted and in the case of wall switches, may be flush or surface mounted on plastic moulded or metal boxes.

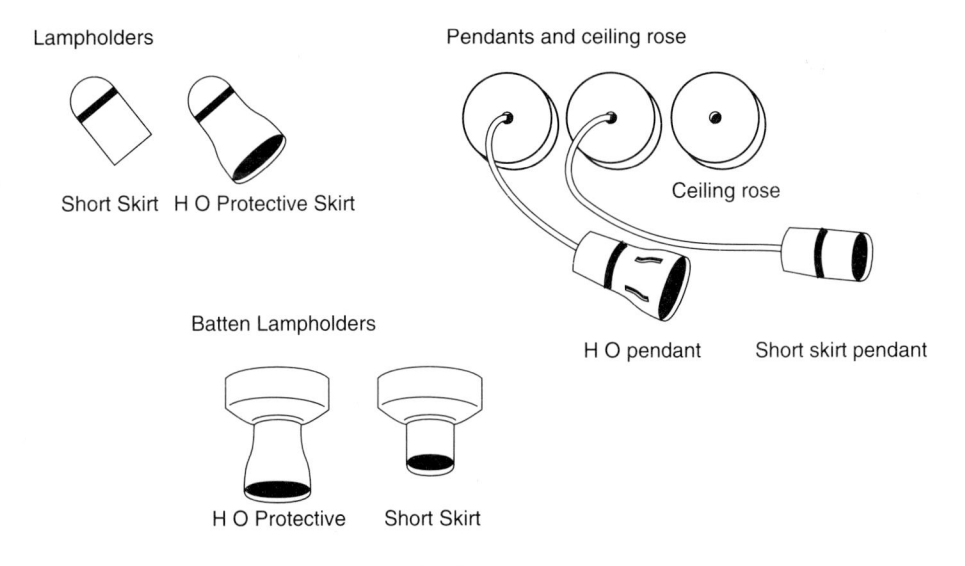

Lampholders

Short Skirt   H O Protective Skirt

Pendants and ceiling rose

Ceiling rose

H O pendant   Short skirt pendant

Batten Lampholders

H O Protective   Short Skirt

Note: HO protective types used in bathrooms.

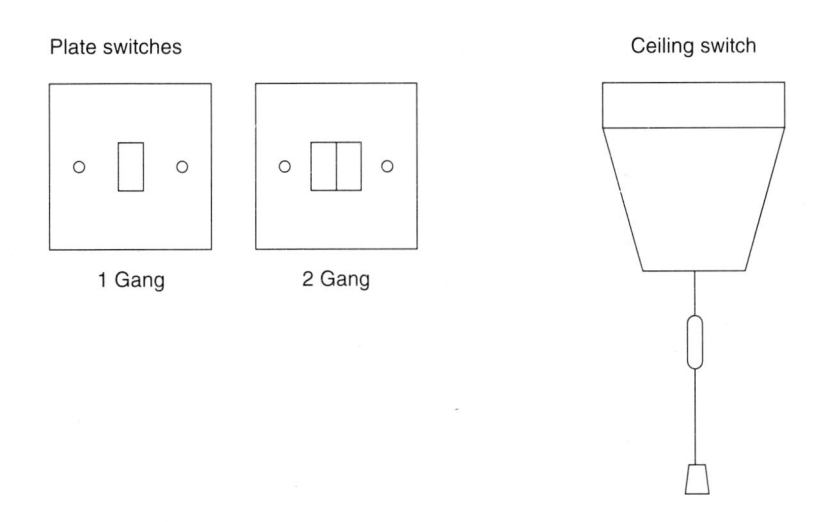

Plate switches

1 Gang   2 Gang

Ceiling switch

For power circuits you will find single or double socket outlets which are available in switched and unswitched formats, again with plastic or metal mounting boxes of the flush or surface type. When control of permanently wired equipment, for example, a central heating boiler or waste disposal unit is required, a switched fused connection unit is required. Switched and unswitched types are available with or without a flex outlet facility.

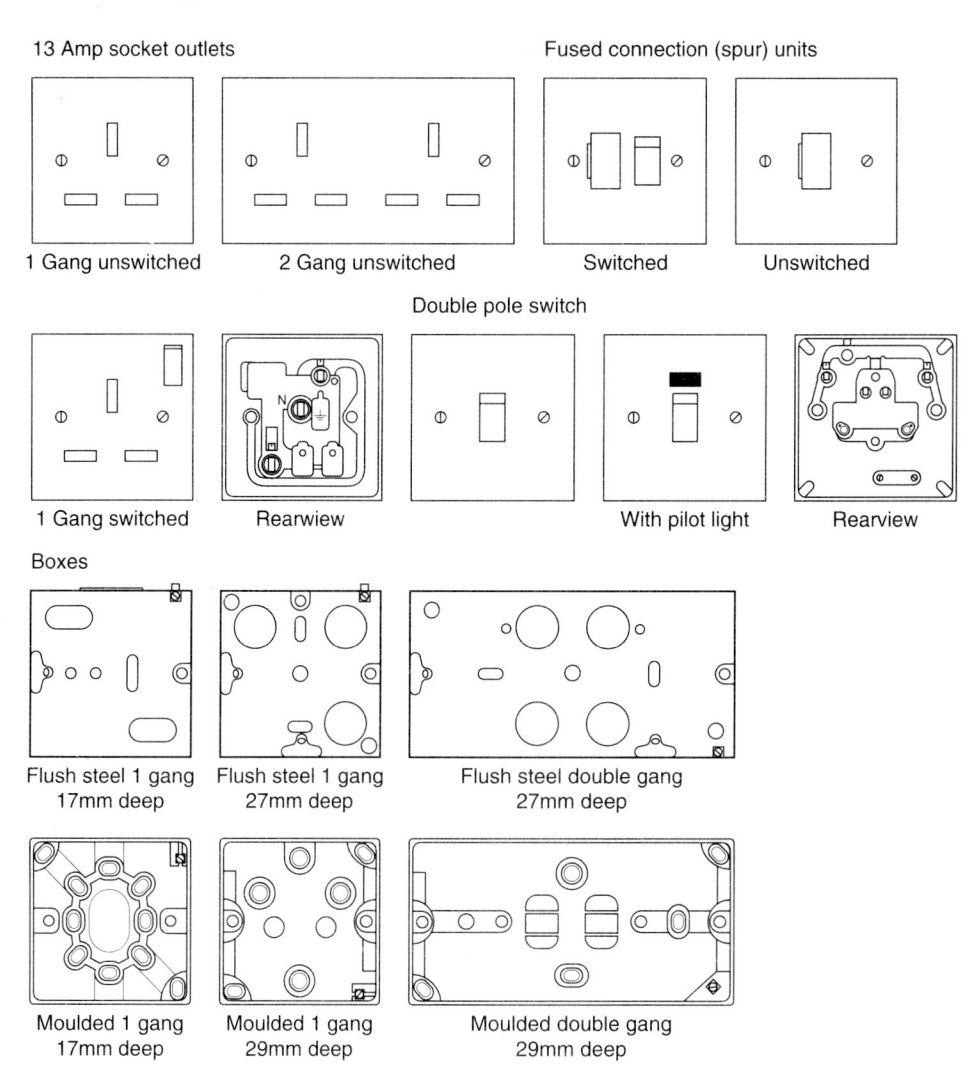

13 Amp socket outlets

Fused connection (spur) units

1 Gang unswitched　　2 Gang unswitched　　Switched　　Unswitched

Double pole switch

1 Gang switched　　Rearview　　　　With pilot light　　Rearview

Boxes

Flush steel 1 gang 17mm deep　　Flush steel 1 gang 27mm deep　　Flush steel double gang 27mm deep

Moulded 1 gang 17mm deep　　Moulded 1 gang 29mm deep　　Moulded double gang 29mm deep

When there is a need to control individual equipment, for example, electric showers and immersion heaters, a double-pole (isolating) switch should be used which disconnects both the phase and neutral conductors from the supply. These switches may be wall or ceiling mounted, with typical current ratings of 20 amp used for immersion heater circuits, 30 or 45 amp used for shower installations with the size being dependent on the rating of the shower.

Double pole ceiling switch

Surface mounting box

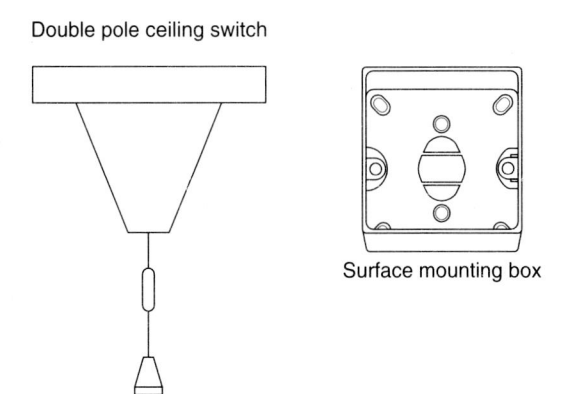

When wiring some lighting and power circuits it is not always convenient to make the connection in the accessory so junction boxes are used. The most common types are 4, 5 and 6 terminal 20 amp and 3 terminal 30 amp.

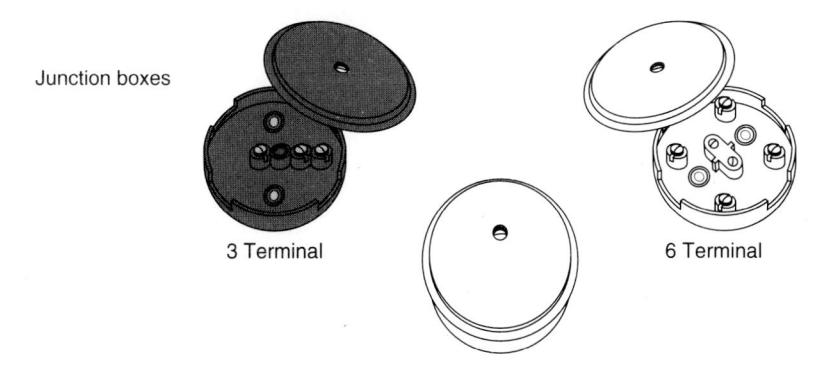

Junction boxes

3 Terminal

6 Terminal

Cables and Flexible Cords

Cables are used to connect the various accessories together to form a circuit.

PVC Cable

PVC (polyvinylchloride) insulated and sheathed cables are used extensively for lighting and heating installations in domestic dwellings, being generally the most economical method of wiring for this class of work.

Types of PVC Cable and Cord

Grades of PVC and their use in cords and cables

PVC compounds used for cords and cables are described in BS 6746: 1969. Several grades of compound are detailed in this standard for both insulation and sheathing requirements. PVC compounds are thermoplastic by nature and consequently are designed to operate within a prescribed temperature range. Grades of PVC can therefore be selected to suit a particular environmental temperature, the maximum conductor temperature for heat resisting grade PVC being 85 degrees centigrade and the lowest operating temperature grade PVC below minus 30 degrees centigrade.

The majority of wiring installations, however, use a general-purpose grade of PVC which has a maximum operating temperature of 70 degrees centigrade, this grade of PVC should not be installed when the air temperature is nearing zero degrees centigrade.

Single Core PVC Insulated Unsheathed Cable

Application.
Designed for drawing into conduit.

Construction

PVC insulated solid or stranded copper conductor coloured red or black. Other colours include blue, green, yellow, white and green/yellow stripes, for use as a circuit protective conductor.

Single Core PVC Insulated and Sheathed Cable with CPC

Application

For domestic and general wiring where a circuit protective conductor is required for all circuits.

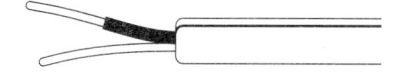

Single Core PVC Insulated and Sheathed Cable

Application.
Suitable for surface wiring where there is little risk of mechanical damage. Single core used for conduit and trunking runs where conditions are onerous.

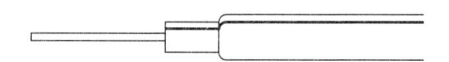

Construction

PVC insulated and PVC sheathed solid or stranded plain copper conductor.
Core colours: black or red.
Sheath colours: black read or grey - other colours available.

PVC Insulated and Sheathed Flexible Cords

Application

General purpose indoors or outdoors in dry or wet locations. Portable tools, washing machines, vacuum cleaners, lawn mowers. Should not be used where the sheath can come into contact with hot surfaces. Not suitable for temperatures below 0°C.

Construction

PVC insulated plain copper conductor laid parallel with an uninsulated plain copper circuit protective conductor, sheathed overall with PVC compound.
Core colour: red. Sheath colour: grey

PVC Insulated and Sheathed Flat Wiring Cables

Application

For domestic and industrial wiring. Suitable for surface wiring where there is little risk of mechanical damage.

Construction

Two or three core cables. Two or three plain copper, solid or stranded conductors insulated with PVC and sheathed overall with PVC.
Core colours: Two core, red and black. Three core, red, yellow and blue. Sheath colours: grey or white.

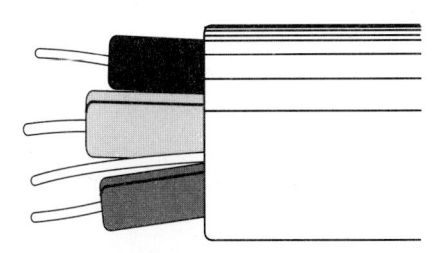

Construction

Two and three core cables exactly as above but with the inclusion of an uninsulated plain copper circuit protective conductor between the cores of twin cables and between the yellow and blue cores of three core cables.

Multicore versions of this cable up to 20 cores have uses in control equipment.

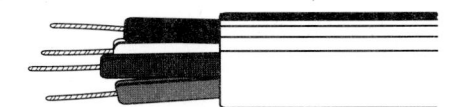

Construction

PVC insulated plain copper flexible conductors laid up and PVC sheathed.
Core colours: brown phase, blue neutral and green/yellow earth, for core combinations up to five cores. Above this, cores are numbered with black numerals on white core insulation.
Sheath colour: black, white or orange.

PVC Insulated and Sheathed Flat Twin Flexible Cord

Application

Intended for light duty indoors, for table lamps, radios and TV sets where the cable may lie on the floor; should not be used with heating appliances.

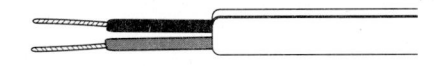

Construction

Plain copper flexible conductors PVC insulated, two cores laid parallel and sheathed overall with PVC.
Core colours: brown and blue. Sheath colour: white.

Heat Resisting PVC Insulated and Sheathed Flexible Cords

Application

Suitable for use in ambient temperatures up to 45°C. Not suitable for use with heating appliances.

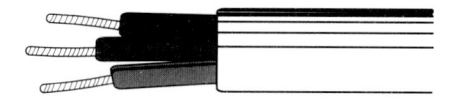

Construction

Plain copper flexible conductors insulated with heat resisting (HR) PVC and HR PVC sheathed.
Core colours: single core, brown or blue; twin, brown and blue; three core, brown, blue and green/yellow; four core, brown, brown, brown and green/yellow. Sheath colour: white.

PVC Insulated and Sheathed Light Duty Flexible Cord

Application

Sometimes known as pendant flexibles, these are used for lighting fittings, push switches and other light domestic applications.

Construction

Two or three core cords having plain copper flexible conductors PVC insulated and sheathed.
Core colours: twin, brown and blue; three core, brown, blue and green/yellow. Sheath colour: white.

Flexible Cords

There are various types of flexible cord. The type required for a particular use depends on the type of equipment it is to be connected to and where it is used.

A guide to the correct selection of a flexible cord for a particular use is given in the Tables of the IEE On-Site Guide to the IEE Regulations.

The size of the flexible cord to be selected is determined by the current rating of the appliance. This is given in Tables in the IEE Wiring Regulations and is based on an ambient temperature of 30°C. Where the temperature is greater than this, correction factors must be applied.

Electrical Cables and Accessories

In addition to the various cables and accessories required for a circuit there is the fuse box, referred to as the consumer unit, which controls the individual circuits such as lighting and power, which together form the complete installation system of a building.

The consumer unit contains the main isolator switch rated at either 80 or 100 amps for isolating the complete installation from the supply. Within the consumer unit you will find an individual fuse for each circuit or an MCB, a mechanical switch, known as a miniature circuit breaker. The purpose of the MCB is to limit the amount of current the circuit it controls can take to prevent cables for overheating under overload or fault conditions. Usually the maximum number fuse of MCB ways in a consumer unit is 12.

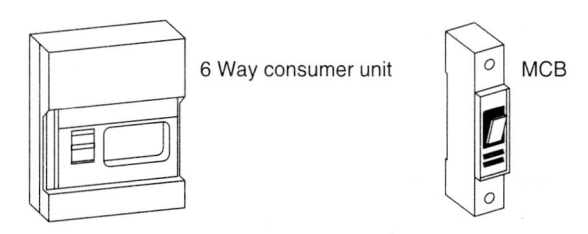

6 Way consumer unit   MCB

On some installations you may also find another protective device called an RCD (residual current device) or an RCCB (residual current circuit breaker). These devices can be fitted in the consumer unit in place of the main isolating switch or be installed near to the consumer unit in an individual enclosure. These devices detect any imbalance in the flow of electricity between the phase and neutral parts of a circuit caused by a leakage of current to earth.

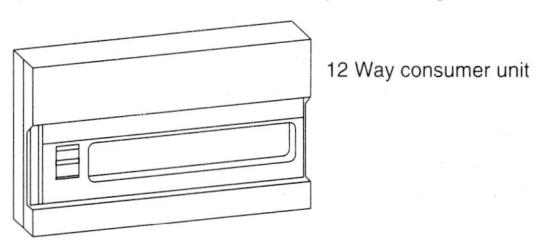

12 Way consumer unit

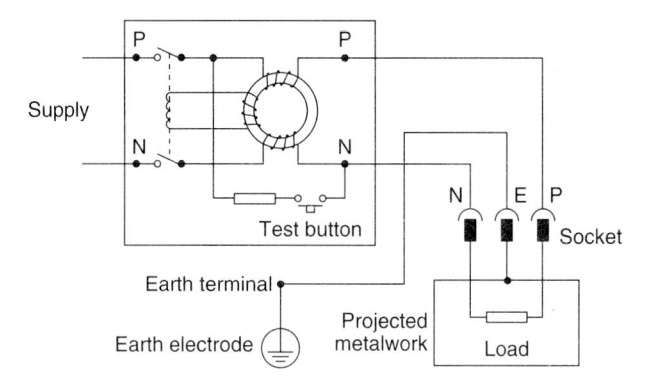

D.P. Switch: connected between incoming supply and load.

Method of Operation Under Fault Conditions

The current taken by the load is fed through two equal and opposing coils wound onto a common transformer core. When the phase and neutral currents are balanced, (as they should be on a healthy circuit) they produce equal and opposing fluxes in the transformer core resulting in no voltage in the trip coil. If more current flows in the phase side than in the neutral side, an out of balance flux will be produced which will be detected by the fault detector coil. The fault detector coil opens the DP switch by energising the trip coil.

Test Switch

The test switch tests only that the circuit breaker is functioning correctly and is operating in the correct order of sensitivity, as specified by BS 4293.

Other Components

In addition to the consumer unit, cable, accessories and other components will be needed to complete an installation in accordance with the standards required by the IEE Wiring Regulations. These include channel, round or flat conduit, metal or plastic, for protective cables buried in walls, mini trunking for cables run off the surface where a neat appearance is required, cable clips for fastening cables to wall and to joists beneath floors and over ceilings.

When metal accessories or boxes are used rubber grommets should be fitted to the knockout holes used for cable entries to prevent chafing of the cable insulations.

In order to comply with the requirements of the IEE Wiring Regulations when a PVC insulated and sheathed cable is terminated to an accessory, the bare earth wire, known as the circuit protective conductor, with a green and yellow P.V.C. sleeving.

Mini trunking

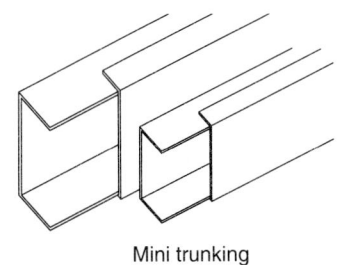

Plugs

In order to connect a portable appliance to a power source a 13 amp plug is generally used which is fitted to the flexible cord of the appliance and is then inserted into the socket outlet. The 13 amp plug has a fuse fitted which is there to protect the appliance and flexible cord against any overload or short circuit conditions. It is very important to fit the correct size fuse, for example a washing machine would require a 13 amp fuse whilst a fan would only require a 3 amp.

The sizes of fuse most commonly available are 3, 5 and 13 amp, whilst 1, 2, 7 and 10 amp types can usually be obtained from manufacturers, to order.

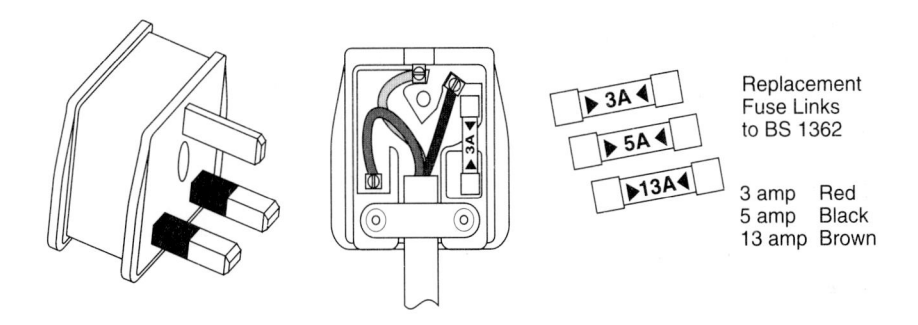

Replacement Fuse Links to BS 1362

3 amp   Red
5 amp   Black
13 amp   Brown

For construction sites, a place where 110 volt portable tools are used, BS 4343 industrial plugs, socket outlets and cable couplers should be used.

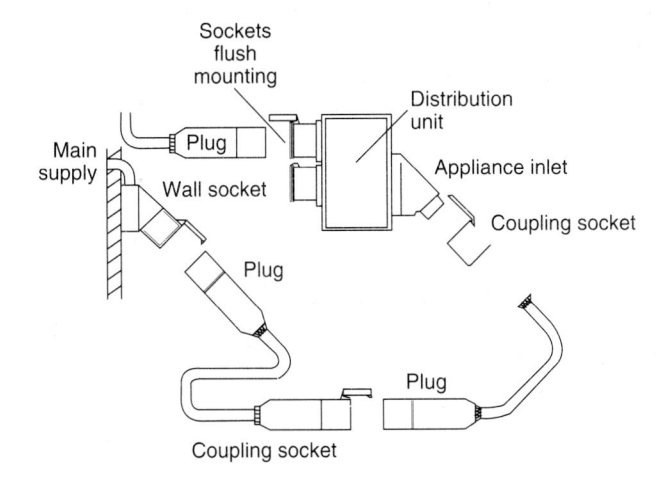

Drawings and Circuit Diagrams

A technical drawing or diagram is simply a means of conveying information more easily or clearly than can be expressed in words. In the electrical industry, drawings and diagrams are used in different forms. Most frequently used are;

>block diagrams,
>circuit diagrams,
>wiring diagrams,
>installation drawings
>cabling diagrams.

Block Diagrams

The various items are represented by a square or rectangle clearly labelled to identify them.

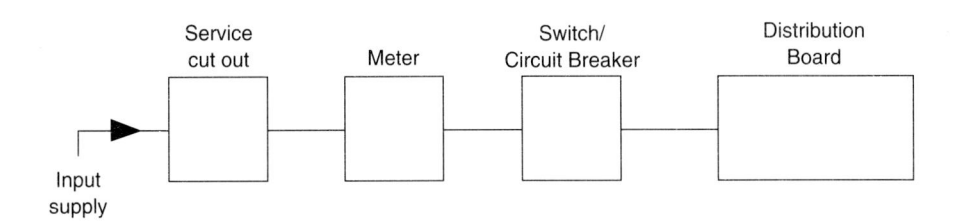

Circuit Diagrams

A circuit diagram makes use of special symbols to represent pieces of equipment or apparatus and to show clearly how a circuit works. It may not indicate the most convenient way of wiring the circuit but it will show the electrical relationship between the various circuit elements.

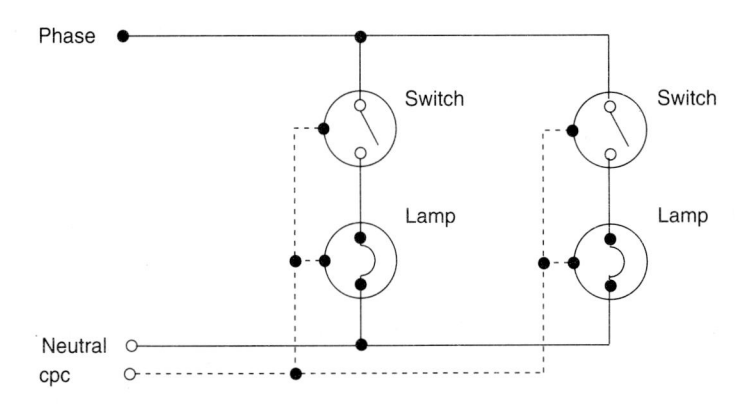

Wiring Diagrams

Wiring diagrams give sufficient information for the connection of a circuit. In some respects wiring diagrams may be more detailed than circuit diagrams but they do not necessarily give any indication of how the equipment concerned operates.

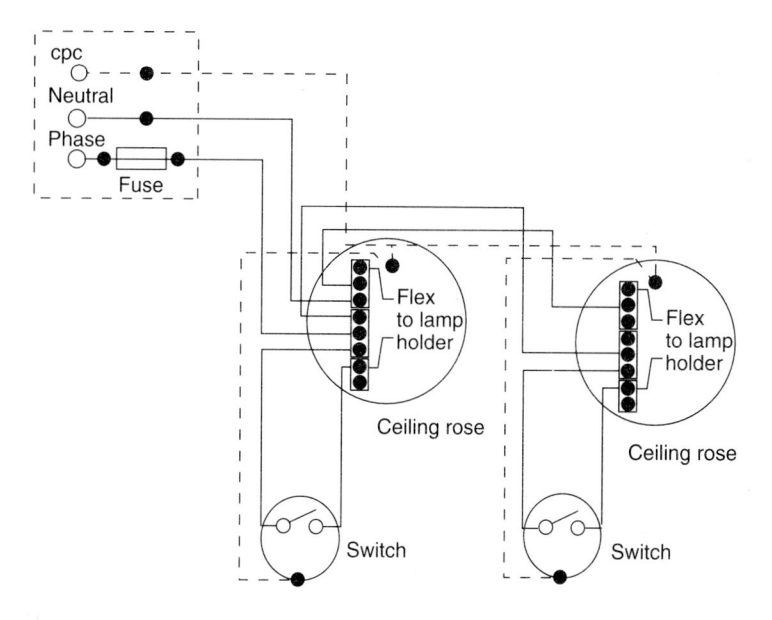

Cabling Diagrams

Show the cables required for the interconnection between components of an installation and will specify the type of cable, number of cores, earth connections etc.

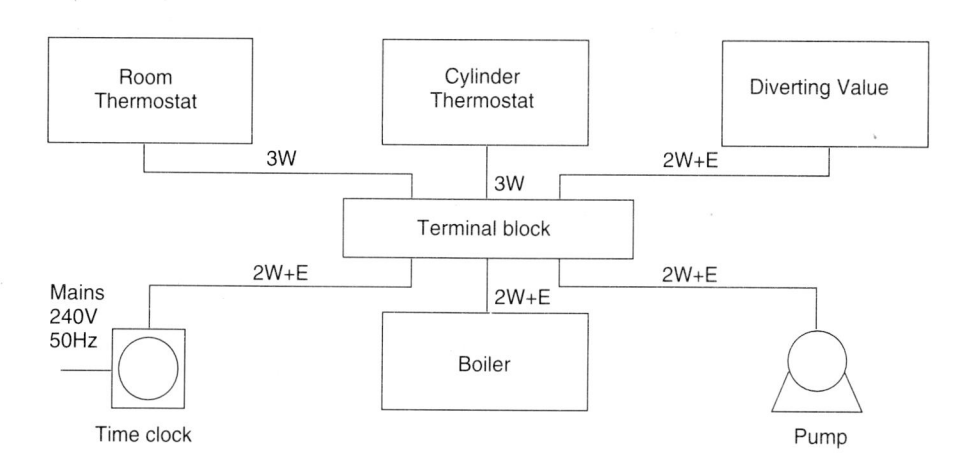

Installation Drawings

These are scale drawings prepared under the supervision of an electrical consultant or engineer responsible for a particular installation and are based upon architects' drawings of the building in which the installation is to be effected. These drawings show the required position of all the equipment, metering and control gear to be installed using symbols to BS 3939.

Symbol	Description
♂	Switch general symbol
	Two-way switch, single pole
	Intermediate switch
	Push button with restricted access (glass cover, etc)
	Luminaire, flourescent lamp general symbol
5	Luminaire with five flourescent tubes
	Self-contained emergency lighting luminaire
×	Lighting outlet position, shown with wiring
⋈	Lighting outlet on wall, shown with wiring running to the left
	Clock general symbol secondary clock
	Buzzer
	Bell

Symbol	Description
	Main control intake point
	Distribution board or point
	Note the circuits controlled by the distribution board may be shown by the addition of an appropriate qualifying symbol or reference. Examples Heating
	Lighting
∞	Ventilating
	Main or sub-main switch
WH	Watt-hour meter
	Water heater, shown with wiring
∞	Fan shown with wiring
	Socket outlet (power) with isolating transformer, for example shaver outlet
	Socket outlet (power) with single pole switch
3	Multiple socket outlet (power) three outlets shown

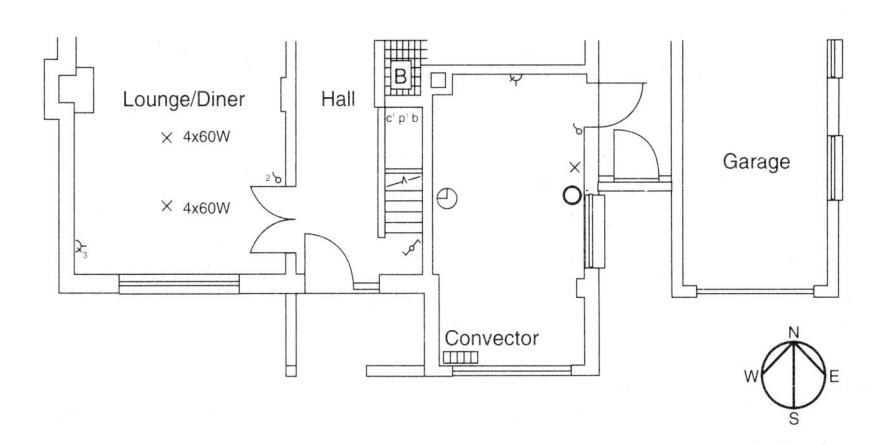

Floor plan showing: Lounge/Diner (× 4x60W, × 4x60W), Hall (c'p'b, B), Garage, Convector, with compass rose (N, S, E, W)

8/3

Lighting Circuits

One-Way Lighting Circuit Diagram

The simplest circuit consists of a pair of wires from the mains terminals supplying a lamp. In this circuit there must be a switch which (if single-pole) must be situated in the phase conductor.

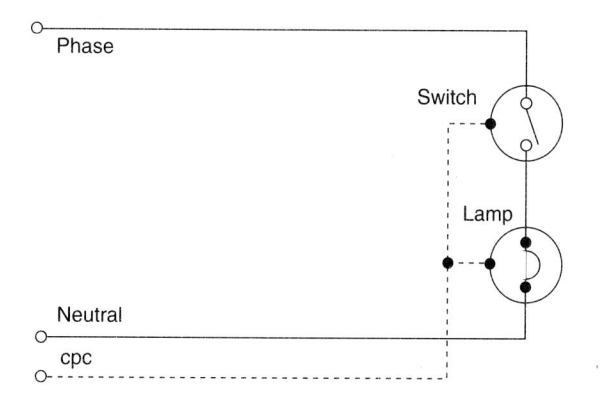

Two-Way Lighting Circuit Diagram

A two-way lighting circuit is often used on staircases so that one can switch off the downstairs light from upstairs, or vice versa.

In this circuit, the switches can have two positions, either of which can light the lamp. Suppose switch A is in the upper position, and switch B is in the lower position, as illustrated, there is no circuit, so the lamp is out.

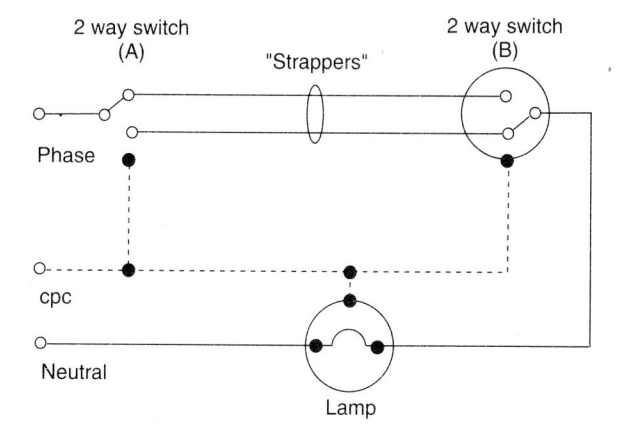

If switch B is operated a circuit is established, and the lamp lights. Now if switch A is operated the lamp goes out. The two wires between switches A and B are called 'strappers'.

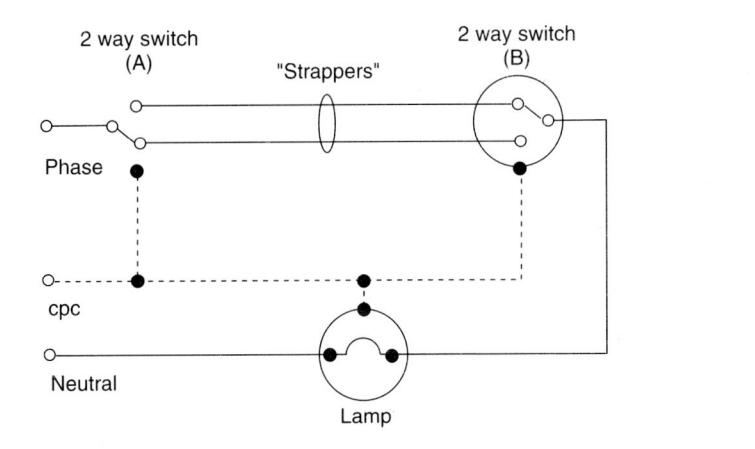

Conversion of one-way light to two-way switching

It is sometimes necessary to modify lighting circuits controlled from one point. Consider an existing one-way circuit such as that previously illustrated. To convert it to a conventional two-way circuit, either the switch feed, or the switch wire must be removed, and replaced by a conductor going to another switch position. This alteration must be made either at the lamp or at the consumer unit.

This can be avoided by using the circuit illustrated. Two-way switch control can be achieved by replacing the original switch by a two-way switch connected as illustrated and by running three new wires to a two-way switch at the new control position.

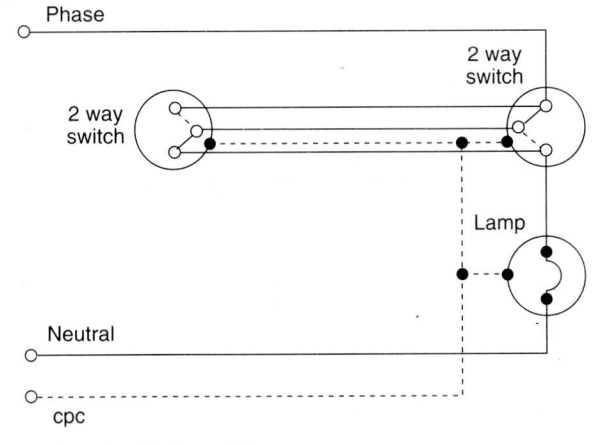

Joint Box Method Wiring Diagram

For two lights independently controlled the circuit would be as illustrated.

1. Joint box showing phase, neutral and circuit protective conductor (c.p.c.)

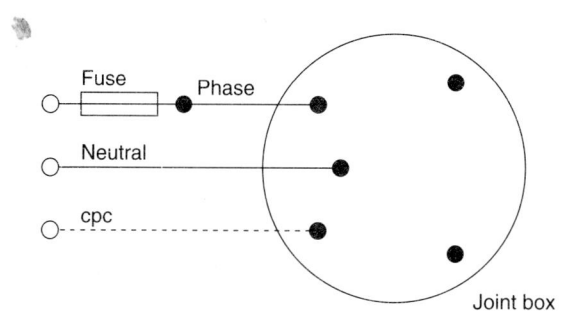

Fuse Phase

Neutral

cpc

Joint box

2. Switch connected showing switch feed and switch wire and circuit protective conductor (c.p.c.)

Switch

Switch feed   Switch wire

Fuse Phase

Neutral

cpc

Joint box

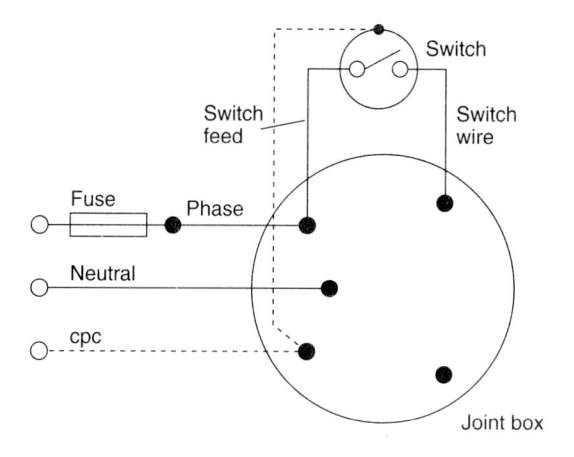

3. Light connected showing switch wire and neutral and circuit protective conductor (c.p.c.)

Switch

Switch feed   Switch wire

Fuse Phase   Switch wire

Neutral   Neutral   Lamp

cpc

Joint box

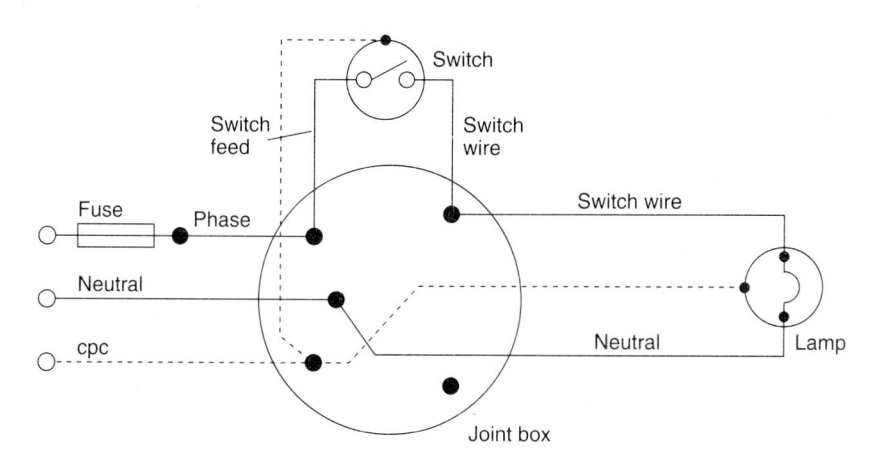

4. Additional light connected showing all switch feeds, switch wires, neutral and circuit protective conductors (c.p.c.)

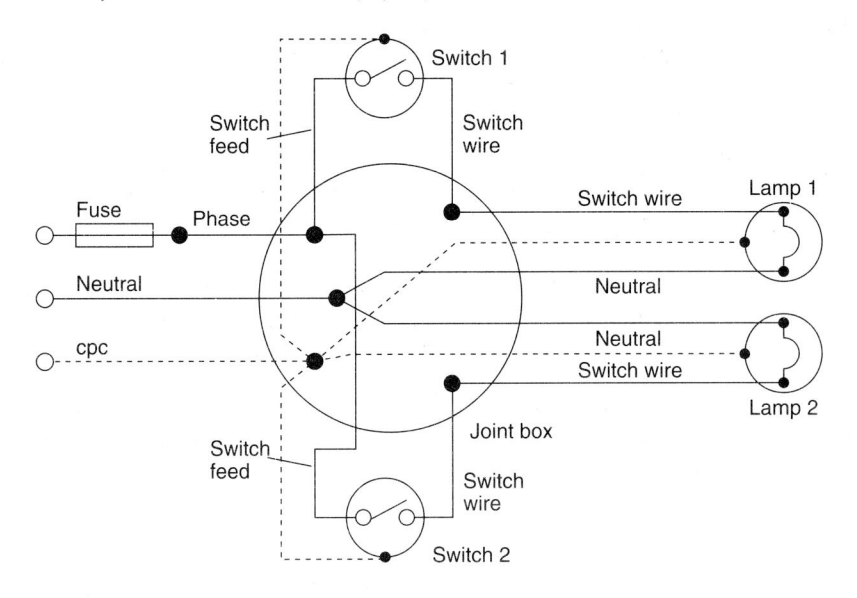

'Loop-in' Method Wiring Diagram

The most common system of wiring final sub-circuits is the loop-in system where all connections are made at the electrical accessories.

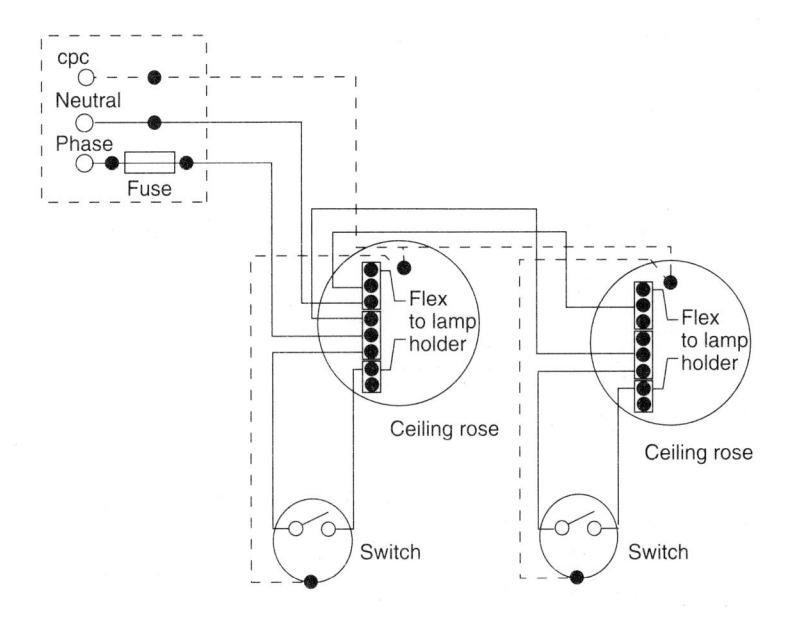

For simplicity all wiring diagrams have to show the basic circuit wiring necessary for the circuit. The loop-in system of wiring at the ceiling rose would be as illustrated.

Number of Points

The number of points which may be supplied by a final circuit is limited by their aggregate demand, this may be assessed using Table 1A of The IEE On-Site Guide to the 16th Edition IEE Wiring Regulations.

Typical examples are:

Lighting Circuits

The current of the connected load or a minimum or 100W per lampholder. For discharge lighting e.g. fluorescent fittings the total load including control gear as supplied by the manufacturer or lamp wattage multiplied by 1.8.

Generally for a single domestic lighting circuit, the number of lights is limited to a maximum of ten.

Power Circuits

For socket outlets other than 2 amp types the rated current of the socket.

Stationary Equipment

The British Standard current rating of the equipment.

Cable Size and Overcurrent Protection

Cable size for wiring domestic lighting circuits is generally 1mm^2 twin with c.p.c. PVC cable; the protective device being limited to a maximum of 6A.

Power Circuits

13A Socket Outlet Circuits

Prior to the early 1950's the multiplicity of types and sizes of socket outlet and plugs in domestic premises had always been a source of annoyance to the householder with 5A and 15A socket outlets installed. In 1947, agreement was reached on a standard socket outlet with a fused plug (BS 1363).

The advantages of this socket are that an appliance can be used in any room; the cost of the wiring is kept as low as possible, whilst affording proper protection for each appliance.

The 13 amp socket outlet system is based on the principle of 'diversity of use'.

Standard Circuit Arrangements

Types of final circuit using BS 1363 socket outlets and fused connection units are:

Ring Circuits
Radial Circuits

The Ring Circuit

In this system the phase, neutral and circuit protective conductors are connected to their respective terminals at the consumer unit and loop into each socket in turn; and then return to their consumer unit terminals, forming a 'ring'. Each 13A socket has two connections back to the mains - each capable of carrying 13A at least.

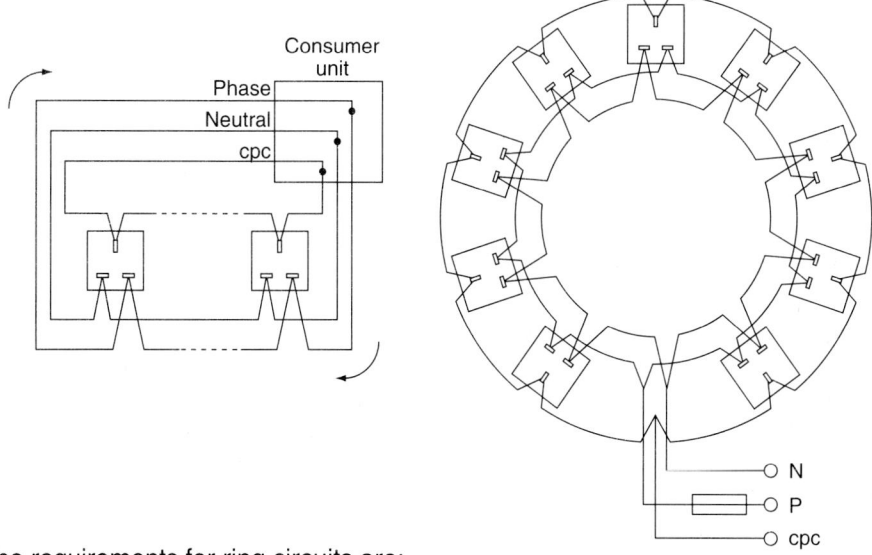

The requirements for ring circuits are:

An unlimited number of socket outlets may be provided. (Each socket outlet of a twin or multiple socket to be regarded as one socket outlet).

The floor area served by a single 30A ring circuit must not exceed 100m^2 in domestic installations.

Consideration must be given to the loading of the circuit especially kitchens which may require a separate circuit.

When more than one ring circuit is installed in the same premises, the socket outlets installed should be reasonably shared amongst the ring circuits so that the assessed load is balanced.

Cable sizes and overcurrent protection are given below. Reference should be made to Table 9A of the IEE On-Site Guide.

Final circuits formats for BS 1363 socket outlets

Circuit	Minimum Conductor Size	Type and Rating of Overcurrent Device	Maximum Floor Area
Ring	2.5mm^2	30 Amp/32 Amp any type	100m^2
Radial	4mm^2	30 Amp/32 Amp MCB or Cartridge fuse	50m^2
Radial	2.5mm^2	20 Amp any type	20m^2

values of conductor size may be reduced for fused spurs.

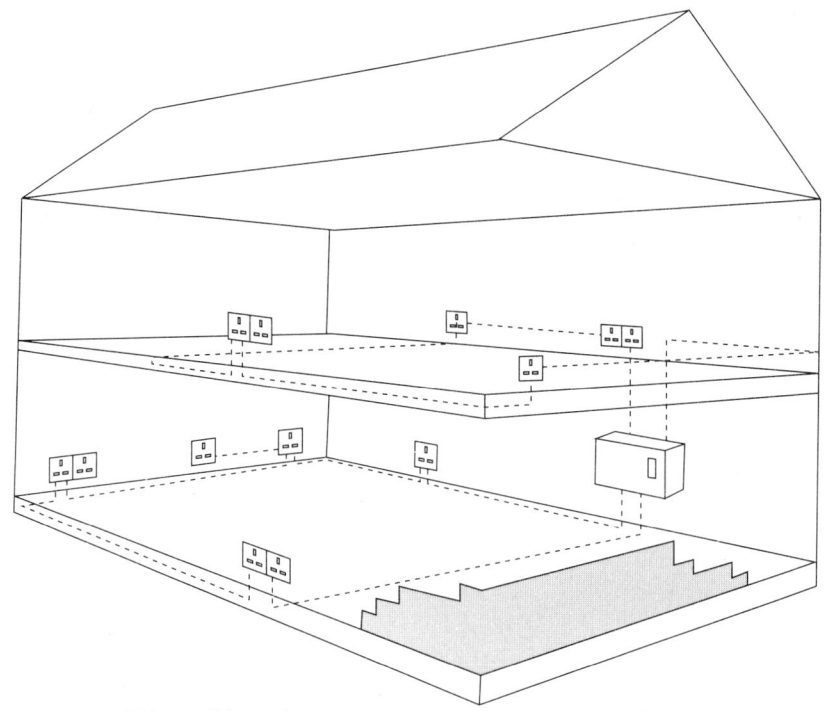

Spurs to a Ring Circuit

Spurs may be installed on a ring circuit. These may be fused – but it is more common to install non-fuse spurs connected directly to the ring circuit conductors. Fuse spurs are connected via a protective device such as a fuse.

The total number of fused spurs is unlimited, but the number of non-fused spurs must not exceed the total number of socket outlets and any stationary equipment connected directly to the circuit. A non-fused spur may supply only one single or one twin socket outlet or one item of permanently connected equipment.

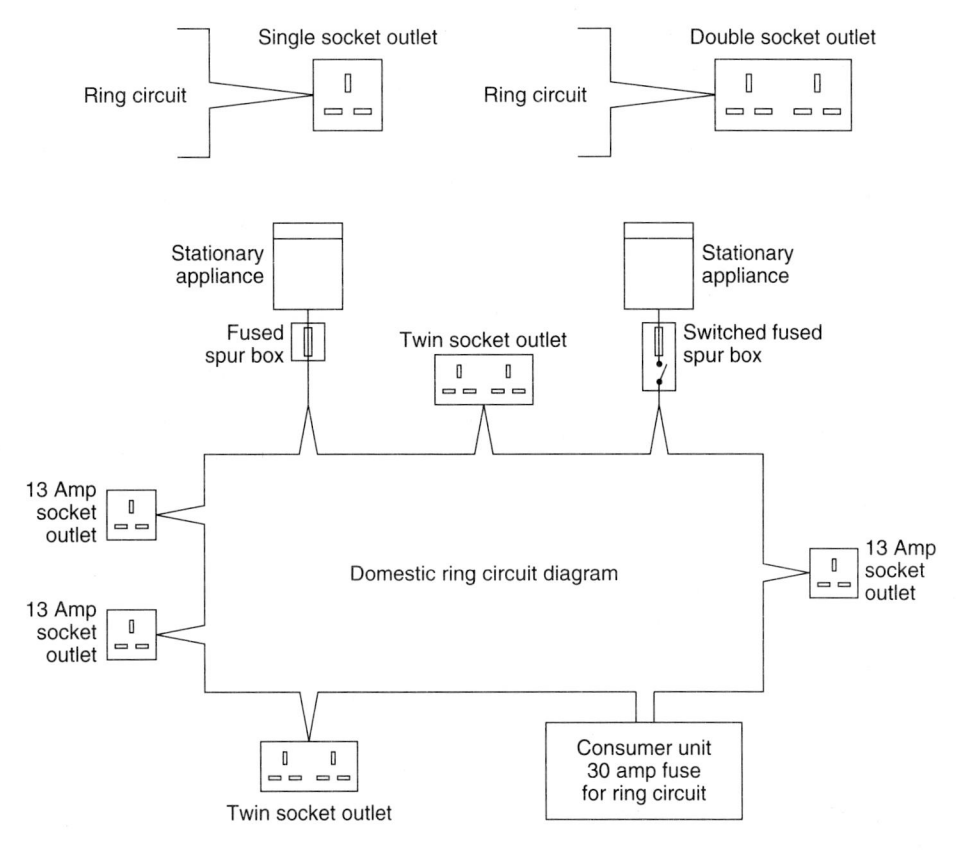

Domestic ring circuit diagram

- Single socket outlet (Ring circuit)
- Double socket outlet (Ring circuit)
- Stationary appliance — Fused spur box
- Twin socket outlet
- Stationary appliance — Switched fused spur box
- 13 Amp socket outlet
- 13 Amp socket outlet
- 13 Amp socket outlet
- Twin socket outlet
- Consumer unit 30 amp fuse for ring circuit

Permanently Connected Equipment

Permanently connected equipment should be locally protected by a fuse (not exceeding 13A) and be controlled by a switch complying with the Regulations, or be protected by a circuit breaker not exceeding 16A rating.

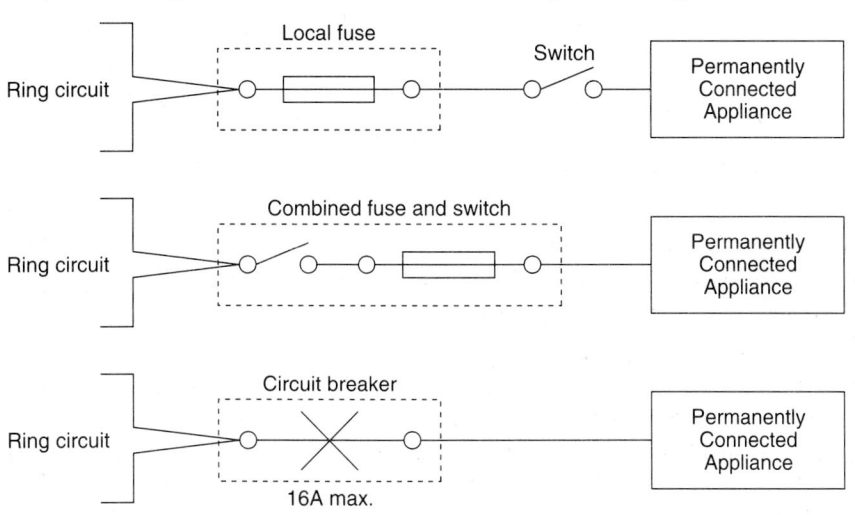

- Ring circuit — Local fuse — Switch — Permanently Connected Appliance
- Ring circuit — Combined fuse and switch — Permanently Connected Appliance
- Ring circuit — Circuit breaker (16A max.) — Permanently Connected Appliance

Note: The cable sizes of spurs should not be less than that of the ring circuit

Fused Spurs

A fused spur is connected to a circuit through a fused connection unit. The fuse incorporated should be relating to the current carrying capacity of the cable used for the spur, but should not exceed 13A.

When socket outlets are wired from a fused spur the minimum size of conductor is:

1.5mm^2 for PVC insulated cables with copper conductors.

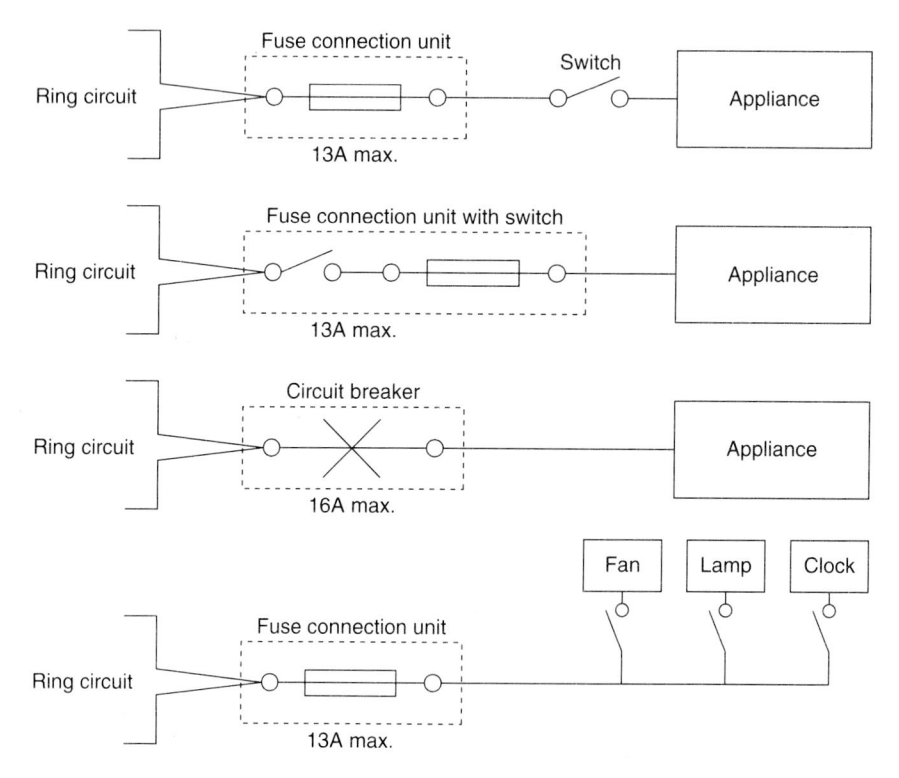

Note: The cable sizes for spurs is dependent on the magnitude of the connected load

(a) at the terminal of accessories on the ring

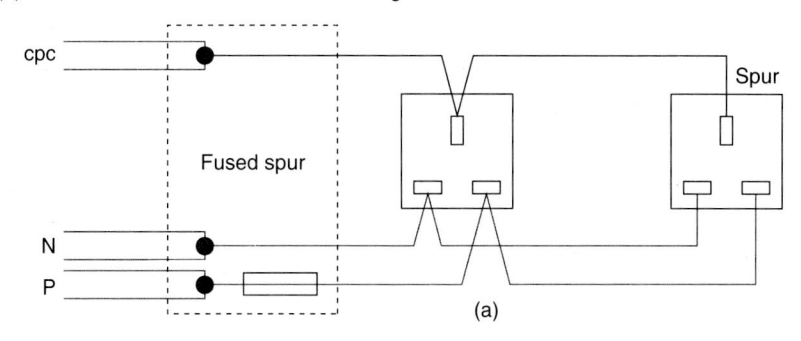

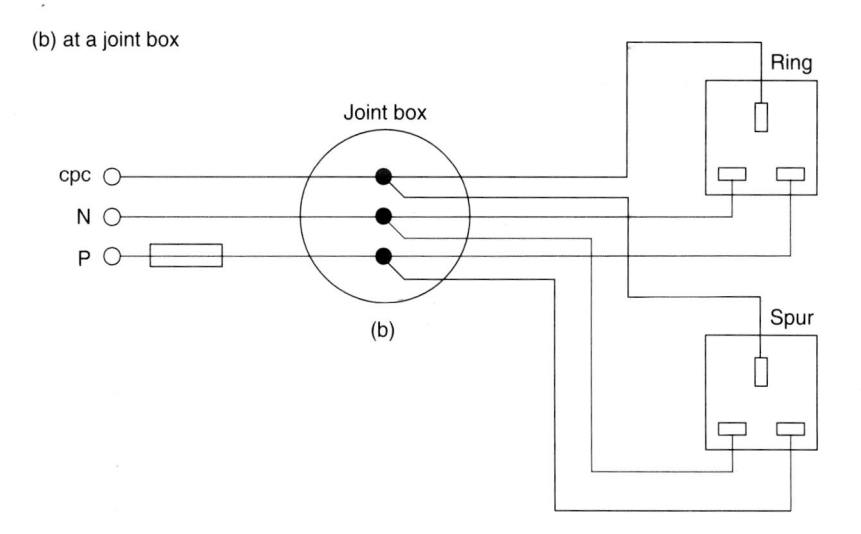

Fused spur

(a)

(b) at a joint box

Joint box

Ring

Spur

(b)

(c) at the origin of the circuit in the distribution board

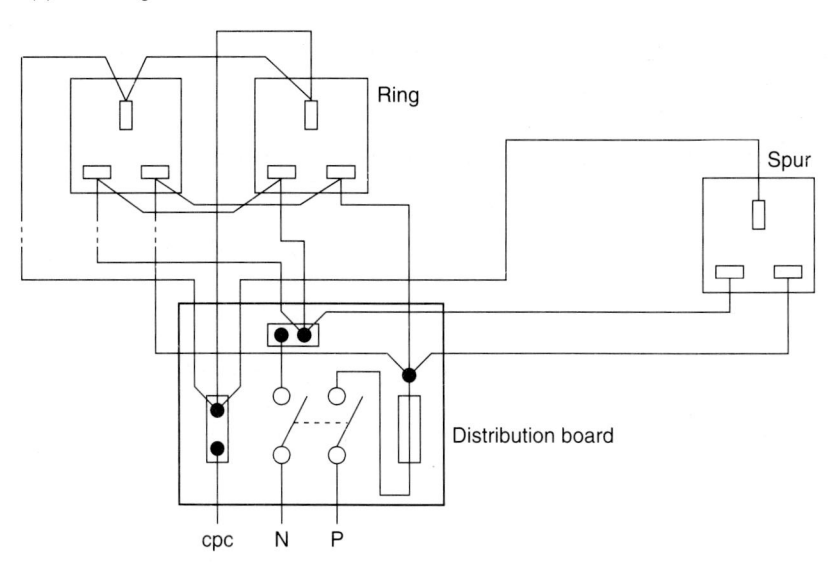

Ring

Spur

Distribution board

cpc N P

Radial Circuits

Radial circuits also make use of 13A sockets (BS 1363) but the circuit is not wired in the form of a ring.

An unlimited number of socket outlets may be supplied, but the floor area which may be served by the socket outlets is limited to either $20m^2$ or $50m^2$ depending on the size and type of cable used and size of overcurrent protection afforded. Refer to Table 9A of the IEE On-Site Guide.

Fixed Equipment

Where immersion heaters are installed in storage tanks with a capacity in excess of 15 litres and electric showers are installed a separate circuit should be provided for each appliance.

Immersion Heaters

Heat resisting flexible cord

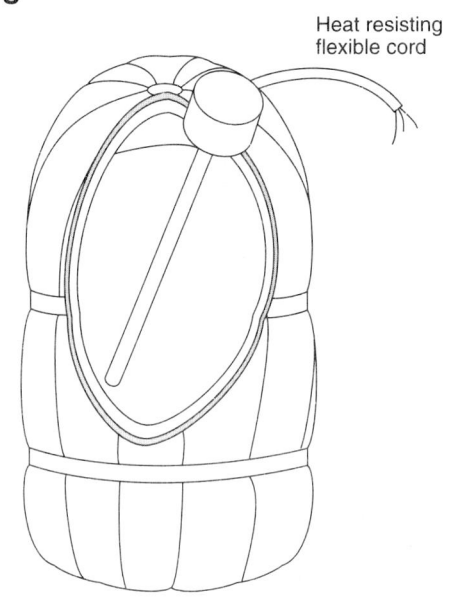

With a permanently installed hot water system, the heater is placed in a cylinder or tank. Hot water is drawn off at sink, basin or bath.
Storage cylinders and tank systems all use electric heating elements immersed in the water, allowing for an efficient transfer of heat. They are generally thermostatically controlled and local isolation is necessary.
The immersion heater must be wired on a separate radial final circuit.
The flexible cord used to make the final connection between the circuit and the immersion heater may reach quite a high temperature, especially if lagging materials cover it. In these circumstances a heat resistant flexible cord should be used.

Immersion Heater with Single Point of Control

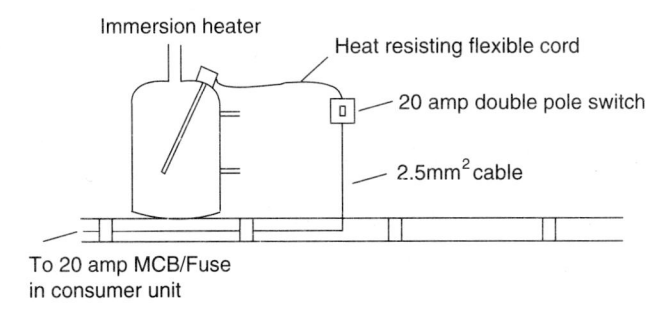

Immersion heater
Heat resisting flexible cord
20 amp double pole switch
2.5mm² cable
To 20 amp MCB/Fuse in consumer unit

Immersion Heater Circuit with Dual Points of Control

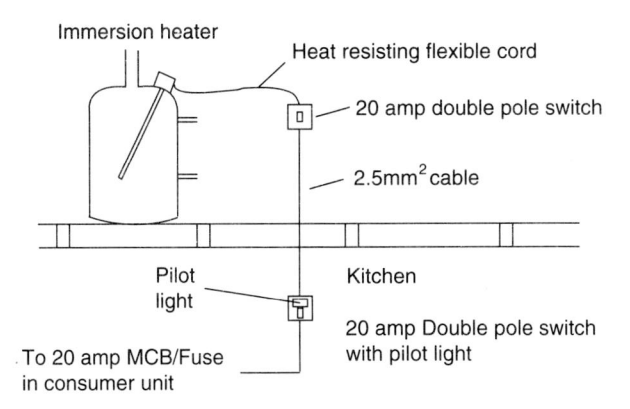

Immersion heater
Heat resisting flexible cord
20 amp double pole switch
2.5mm² cable
Pilot light
Kitchen
20 amp Double pole switch with pilot light
To 20 amp MCB/Fuse in consumer unit

A typical final circuit diagram is illustrated below.

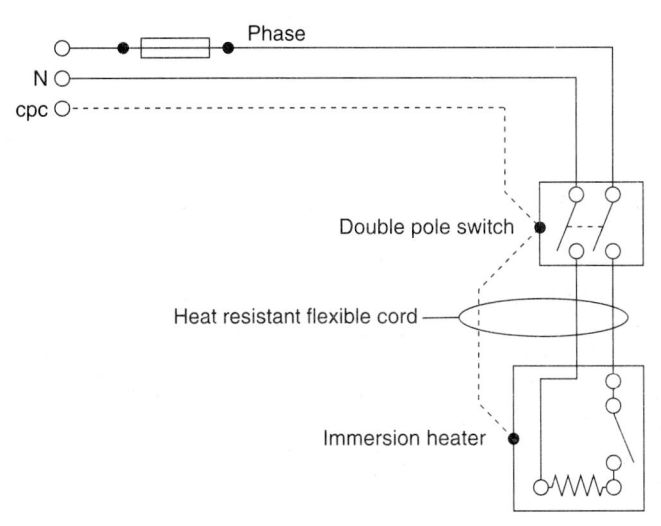

Phase
N
cpc
Double pole switch
Heat resistant flexible cord
Immersion heater

Dual Immersion Heater Circuits

It is possible to fit two immersion heaters with thermostats into the same cylinder to achieve either a full or a part cylinder of hot water.

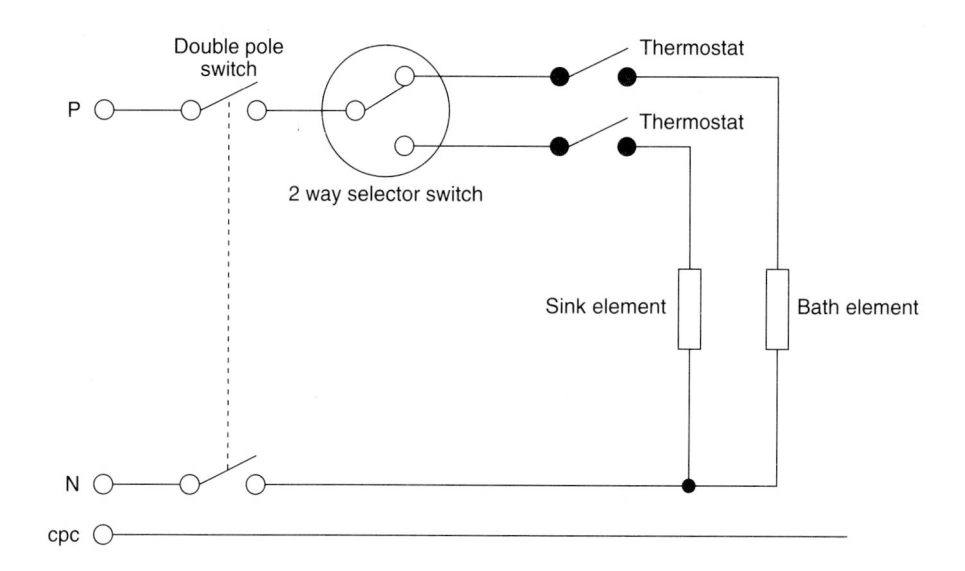

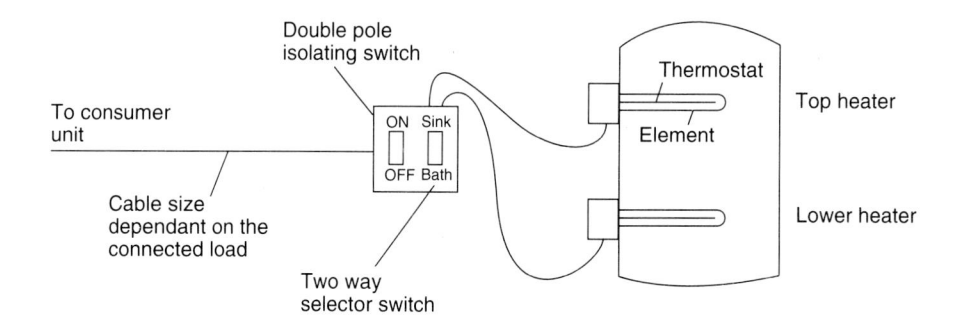

Tools and Installation Practices

The Electrician's Tools

It is difficult to list all the tools an electrician might need in his work, including those with specialised uses. This section describes those most commonly used and should not be regarded as exhaustive. Much of the information given here is elementary, but is included for the benefit of persons who may not have had the opportunity to familiarise themselves with the range of tools in common use.

Hammers

Hammers have three main functions; to drive another tool, to drive a fixing, or to apply force to the workpiece directly - as in shaping metal or in demolition. It is important that the tool is suited to the purpose it is being used for. The main types of hammer likely to be used by an electrician are described below:

Ball Pein Hammer

Available with a variety of headweights up to about 3lbs, this is primarily an engineer's hammer; the ball pein being used for riveting and shaping metal. The heavier ball pein hammers are suitable for driving cold chisels and punches.

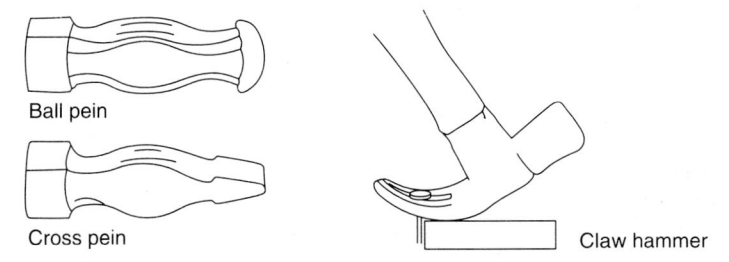

Ball pein

Cross pein

Claw hammer

Cross Pein Hammer

Primarily a carpenters hammer, its tapered pein is used to start nails held between the fingers. General use. Head weights up to 18ozs.

Claw Hammer

For general use. The curved claw is used to lever and pull nails out. This function demands that the head be securely attached to the shaft. With steel shafted hammers, the head is permanently fixed to the shaft; hickory shafts are secured with wedges driven in to spread the shaft inside the eye. Work surfaces should be protected by placing a piece of hardboard under the head of the hammer when it is used as a fulcrum for levering.

Lump Hammer

A heavy duty double faced hammer suitable for demolishing masonry, driving cold chisels and masonry (star) drills.

Saws

The hacksaw and the tenon saw are probably most widely used by electricians. However, a few other saws with specialised uses are equally important.

Hacksaw

Basic metal working saw for cutting off, making thin cut and removing surplus material. Consists of a frame, handle and blade. The blade is tensioned by a wing nut or sometimes by the spring of the saw frame itself.

Tenon Saw

Intended primarily for cutting joints and small section timber to length. The blade has parallel edges top and bottom with a reinforcing metal strip along the back to give rigidity. The handle is mounted high on the blade. In use, once the cut has been started, the saw will usually be moved square to the work (i.e. held almost level) when cutting joints.

Padsaw, Keyhole Saw

These saws have no frame and consist of a narrow tapered blade fixed to a handle, so that different sizes of blade can be used on one saw. Since the blade consists of little more than a narrow strip of steel, the saw can be worked in a very limited space; for example, once the material has been drilled to receive the blade, it can be used to cut a small hole in the centre of a panel.

Flooring Saw

Used for cutting tongue and groove floorboards prior to lifting them. The saw has a convex blade which allows the tongue to be cut away from the floorboard when it is still in place.

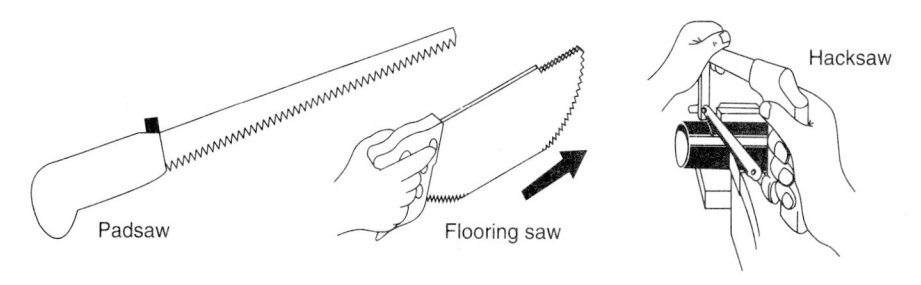

Chisels and Punches

Cold Chisel

A heavy chisel used for rough cutting metal, 'flat chipping' along the surface of the workpiece, and removing rivet and bolt heads. Generally used with a ball pein hammer.

Bolster Chisel (Feather Chisel)

A chisel with a spade shaped blade, the widest part forming the cutting head. Used for prising up floorboards. The width of the blade allows considerable force to be applied once the bolster is securely inserted in the gap between boards.

Wood Chisel

Used to trim wood by paring, cutting and clearing waste from joints. May be driven by a mallet, or by the heel of the hand when less force is required.

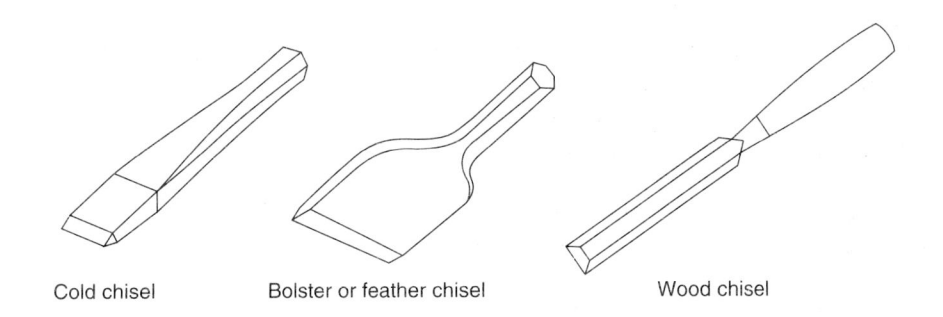

Cold chisel   Bolster or feather chisel   Wood chisel

Centre Punch

This tool consists of a knurled shaft (for holding between the fingers) with a sharp point at one end. The head of the punch is struck with a hammer. Centre punches are used to produce an indentation in metal for starting a drill accurately, and also for accentuating marked lines on metal work with a series of indentations.

Nail Set or Punch

Similar in appearance to the centre punch, the nail set has a flat ground tip for sinking nails below the surface of wood, without causing damage to the surface with the hammer head. Centre punches should not be used for this purpose.

Centre punch

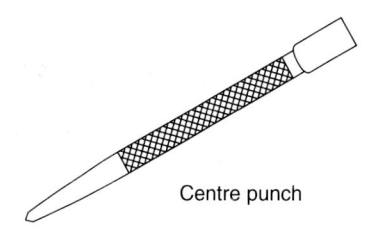

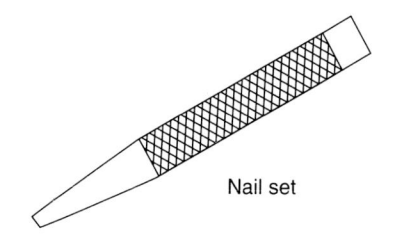

Nail set

Drills

The drill is used to bore holes in solid material. A mechanism, which may be hand or power driven, applies a continuous turning motion to the drill bit causing the cutting edge at its tip to bore into the material being worked.

The main types of drilling machine are the wheelbrace (or handrill) rotated by means of a wheel, the breast drill, similar in principle, but applied to the workpiece while bracing the shaped end-plate against the body, and the carpenter's brace which is worked by rotating the frame while applying pressure to the handle at the head of the brace. The main types of power drill are the hand held electric drill, and the drill press or bench drill, which is a workshop machine. Drill bits are available in a wide range of types and sizes suited to a variety of purposes. The drill bit should generally be suited to the drilling speed and the material being drilled.

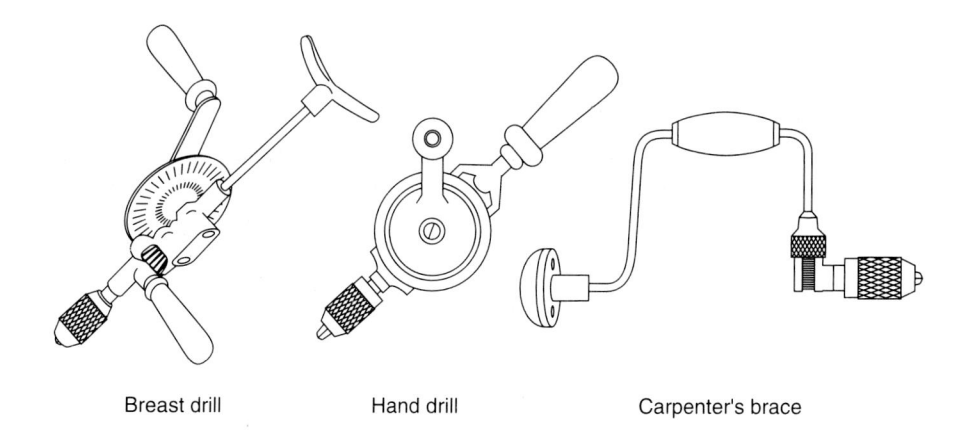

Breast drill

Hand drill

Carpenter's brace

Rawlplug Holder and Bits

A percussion hole boring tool, driven by a hammer. Consists of the rawldrill and the toolholder into which it fits. The tool is used to make holes in masonry for plug fixings.

Stardrill

A hand held boring tool for use on masonry. Resembles a cold chisel; but has a fluted shaft tipped with four ground cutting faces. It is held in position and struck with hammer, (while being turned between each blow) to produce a hole in the material.

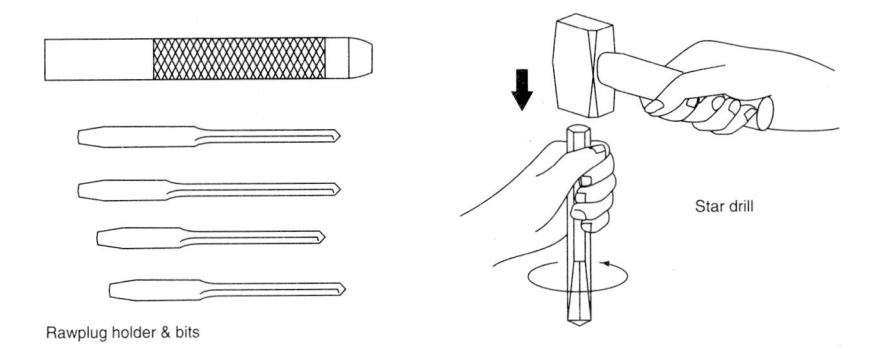

Rawplug holder & bits

Star drill

Bradawl

Used to start holes in timber for the entry of screw fixings. Blades with various tips are available - the most widely used probably being the 'screwdriver' type. It should be entered into the wood across the grain, and twisted in alternate directions.

Screwdrivers

A wide variety of types and sizes, some having very specialist uses, are available. There are three main types of screwdriver tip in common use. The flare tip (designed to fit the familiar slotted screw) has a flat tapered tip which flares out from the blade. Parallel tips are tapered but not flared - so that the tip is no wider than the blade of the screwdriver, and can be used to drive screws recessed into holes. They are used on slotted head screws. The cross-head tip (Phillips, Posidrive) has four flutes ground on the head, formed to a point, which fits the corresponding crossed slots in the screw head. It has the advantage of giving extra purchase on the screw.

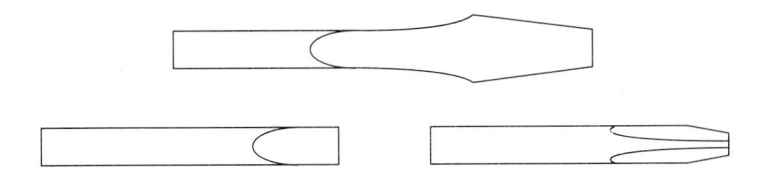

Electrician's Screwdriver

This has an insulated fluted plastic handle, with a long thin blade (sometimes also insulated) with a parallel ground tip. They are available in a variety of sizes.

Cabinet Screwdriver

The standard woodworker's screwdriver with bulbous handle designed to fit into the palm of the hand. The tip is flared, sometimes with the points ground back to narrow the head.

Wrenches and Spanners

These are used for tightening nuts, bolts and set screws, and 'span' the head to provide leverage to turn the nut or bolt as required. Spanners are usually marked with the size of the nut they are to fit. An exception to this rule is the BSF sizes where the spanner size fits the nut a size larger than the nominal.

Wrench (Footprint)

The name is related to the trade mark of the manufacturer. The jaws of the wrench are serrated and are adjustable to a wide range of sizes.

Wrench (Stilson)

This is a more powerful version of the Footprint wrench. Its jaws are also serrated and it tightens on the nut.

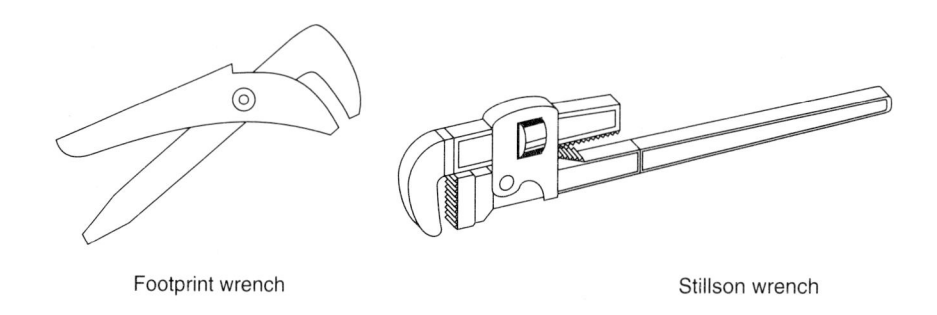

Footprint wrench

Stillson wrench

Adjustable Wrenches and Spanners

These can be adjusted to suit the size of nut or bolt. They give less positive grip than is obtained with an open ended or ring spanner and should only be used when the correct size of spanner is not available.

Vice Grip

These are also called mole grips, and can be pre-set with the adjustment screw to lock on to plates of given thickness. They enable the work to be held securely. They are released by a trigger.

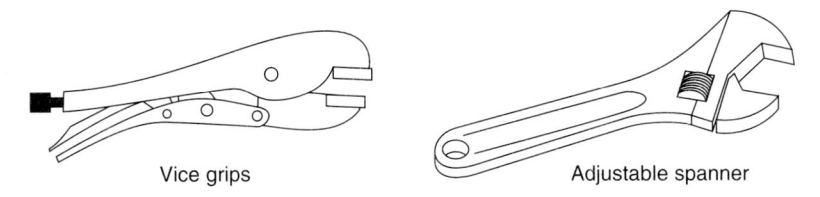

Vice grips

Adjustable spanner

Open Ended Spanner

The type of spanner commonly used is the 'open' type, either single or double headed, with the opening fashioned so that it encloses three sides of a square, or four sides of a hexagonal nut. The advantage of this type of spanner is that it may be applied at the side and does not require to be placed in position over the nut or bolt head. The sides of the jaws may be parallel to or inclined at an angle (usually 15 degrees) to the handle. The incline shape is useful when the nut or bolt is in a position where only small movements of the spanner are possible.

Ring and Box Spanners

With ring and box spanners there is less danger of the spanner slipping since they surround the nut and the risk of spoiling the nut is greatly reduced. Box spanners can be used in positions where the application or movement of other types of spanner is impossible. The shape of the ring or box may be square or hexagonal, or the ring may be in the form known as bi-hexagonal, which has 12 points thus giving the advantage of additional spanner positions.

Allen Key

An L-shaped bar which fits into the hexagonal recess in the head of a machine screw or bolt enabling it to be turned.

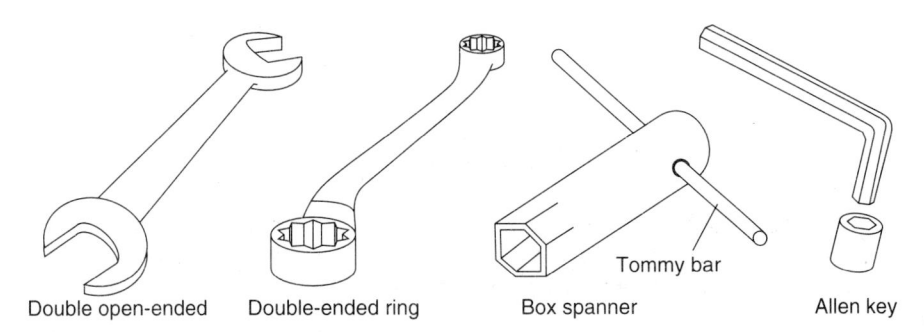

Double open-ended   Double-ended ring   Box spanner   Tommy bar   Allen key

Pliers and Cutters

Electrician's Pliers (Combination Pliers)

These are basically engineer's pliers fitted with insulated handles. They have serrated jaws for gripping and bending, behind which is usually a curved serrated section for gripping metal rods, and nearest the pivot, a pair of side cutters for cropping wire.

Long, Round, Snipe, Nose Pliers etc.

Long nosed pliers have long, tapered jaws suitable for gripping in confined spaces. Bent, snipe and needle-nosed pliers have similar specialised functions. Round nose pliers are used to form loops in wire prepared for termination to an electrical fitting.

Insulation Strippers

Resembling pliers, wire strippers have the tips of their jaws turned inwards at right angles towards each other. When closed to a pre-adjusted setting over a cable, the cutting tips sheer off the insulation from the core, leaving it prepared for termination.

Side Cutting Pliers

Designed only for cropping, these pliers should not be used for other purposes. The jaws are angled, allowing the handles to be held away from the work surface to provide knuckle clearance.

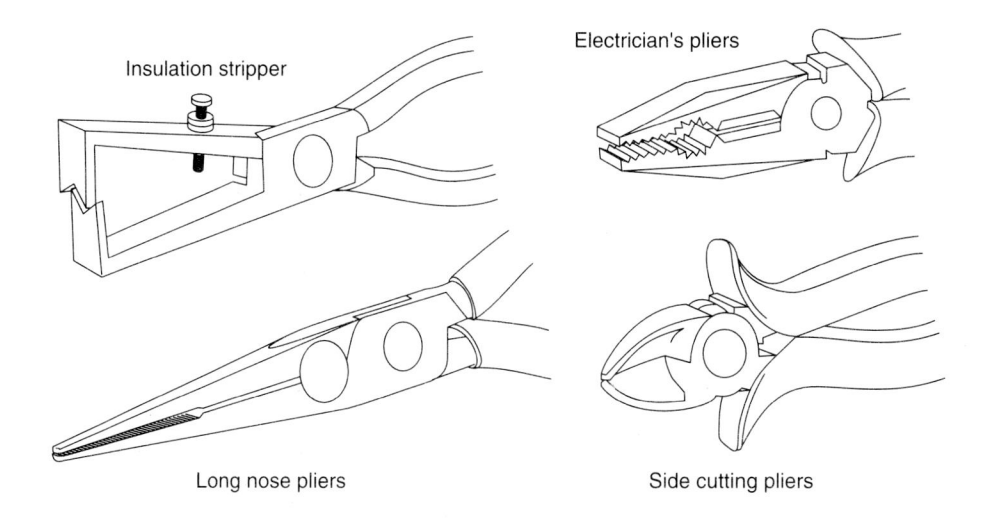

Insulation stripper

Electrician's pliers

Long nose pliers

Side cutting pliers

Electrician's Knife

Designed for electricians. Size and blade designed for the stripping of PVC cable. Usually hand finished and made from the finest Sheffield steel. Length of knife 3 1/2".

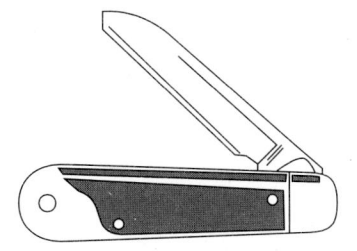

Measuring and Marking Off Tools

Steel Tape

A steel flexible strip rule coiled into its container and spring loaded. Up to 5m long. Compact and easily carried in the pocket. Useful for measuring in restricted spaces and awkward locations.

Spirit Level

Essentially a straight edge in which is set a glass tube filled with liquid and enclosing an air bubble. If the surface against which the straight edge is set is level, the bubble will centre itself between two lines on the tube.

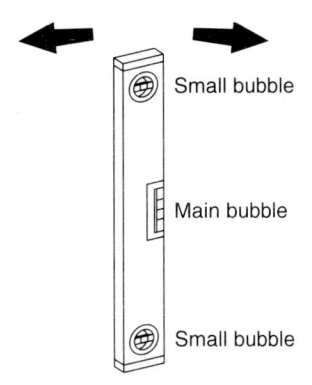

Plumb Line or Plumb Bob

Used to find the vertical when marking off walls etc. Consists of a weight (plumb bob) attached to a line. The line is held up to a marked point from which the vertical is to commence. The weighted end is allowed to swing free to the desired length; when it steadies, the true vertical will be indicated.

Chalk Line

Essentially a line coated with chalk which is strained taut between two pre-determined points, then plucked back and allowed to snap to against the surface, where it will deposit a line of chalk. It is mainly used to mark long lines.

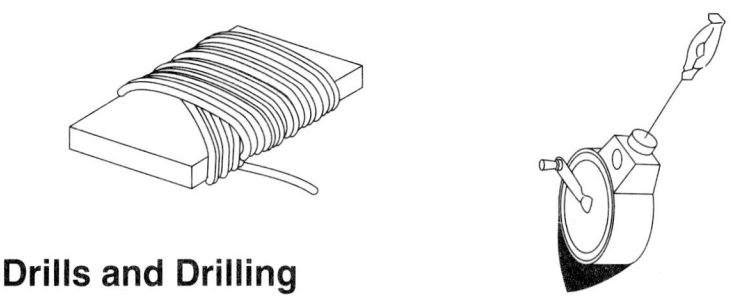

Drills and Drilling

During the course of his work the Electrician will be required to drill holes in a wide range of materials: metals, wood, masonry, etc. and will need to use a variety of drills and drill bits.

Perhaps the most common tool in use is the hand-held portable electric drill used with carbon steel or high speed bits (twist drills) for drilling small holes into metal, wood and partition materials and with Durium tipped drill bits for use in drilling into brick or stone. Where no electric power is conveniently available, a wheel brace or breast drill is used and for masonry, a Rawldrill.

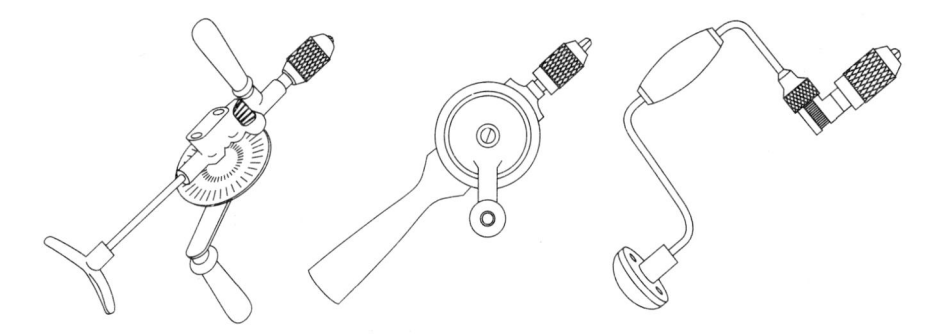

In cases where a large diameter hole is needed, a carpenters brace and bit is commonly used for holes in wood and partition materials, and star drills or Rawlcore drills for masonry.

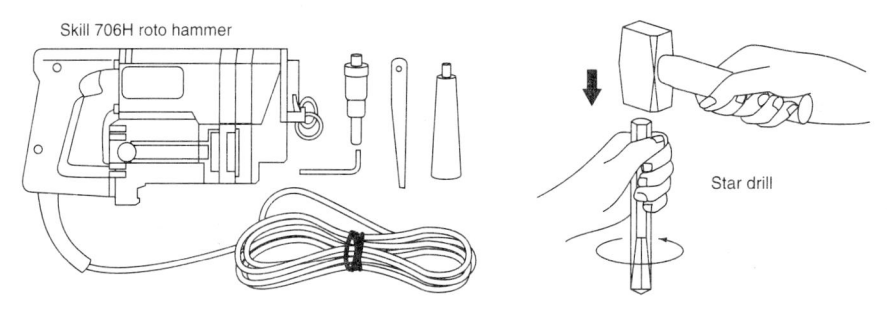

Skill 706H roto hammer

Star drill

Rotary/impact or hammer drills are often available on building sites and are particularly useful for drilling holes into hard masonry. They are used with toughened steel bits having specially hard carbide tips, designed to withstand rapid percussion.

Power Tools

For installation work in buildings where an electrical supply is available, 110 volt electric drills are required for installation in new building construction work or repairs, alterations, extension of existing buildings.

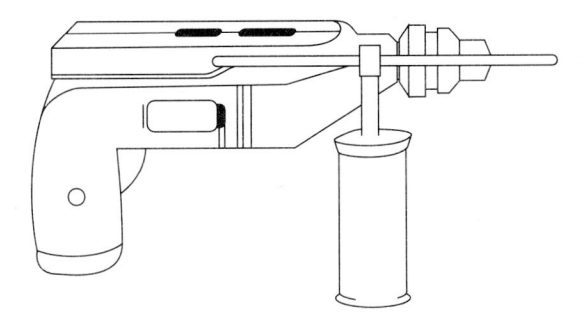

These drills often come with a hammer feature which can be accessed by pressing a button or turning a knob.

The hammer feature enables the drilling into masonry material much faster as the drill bit is hammered into the material as it is rotated.

Cordless Tools

For installation work where an electricity supply is not readily available cordless electric drills can be used.

For any situation, the cordless electric screwdriver may prove to be an advantage over the hand held screwdriver.

Tipped Masonry Drills

Rotary masonry drills may be tipped with Durium, a specially hard abrasion resistant carbide, or with Tungsten carbide. The fluting design ensures quick spoil removal resulting in greater drilling accuracy and longer drilling life.

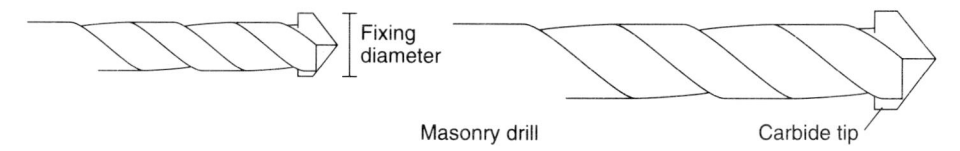

Masonry drill — Fixing diameter — Carbide tip

Drill sizes range in cutting diameter from 4mm - 25mm. Longer drills are available to solve the problems of extra deep drilling.

Tipped masonry drills will penetrate easily into brickwork and marble, as well as glazed tiles, slate, hollow building blocks and concrete.

The longer drills are ideal for penetrating thick walls for conduit, cables, plumbing etc. but are expensive and should be used with care.

Method of Use

1. For best results use at low speed in a rotary drill or a wheelbrace. Use eye protection when drilling.
2. Exert a firm pressure behind the drill for fast accurate penetration.

Rotary Impact Masonry Drills

These are specially manufactured for use in rotary impact drilling machines, and have a hard carbide tip, designed to withstand rapid percussion. The tip is supported by a high quality tough steel shank.

The standard series of drills include drills ranging in cutting diameter from 4mm - 25mm. Longer drills are available to deal with the problem of extra deep hole boring.

The drills will bore rapidly and accurately into structural concrete, engineering brick and other hard masonry

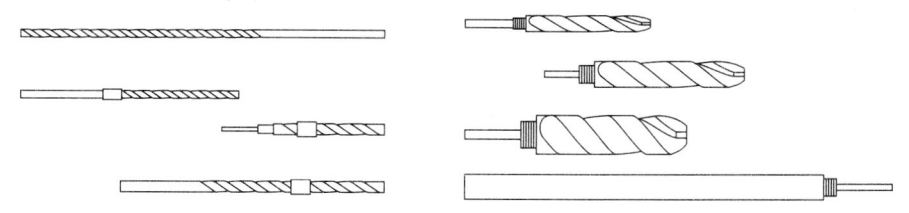

Method of Use

1. Use in a rotary impact drilling machine. Use eye protection when drilling.
2. Exert a firm pressure behind the drill for fast accurate penetration.

Hammer Drills

Hammer drills are designed for use with the wide range of rotary hammers available. These machines have a more powerful hammer action than rotary impact drilling machines.

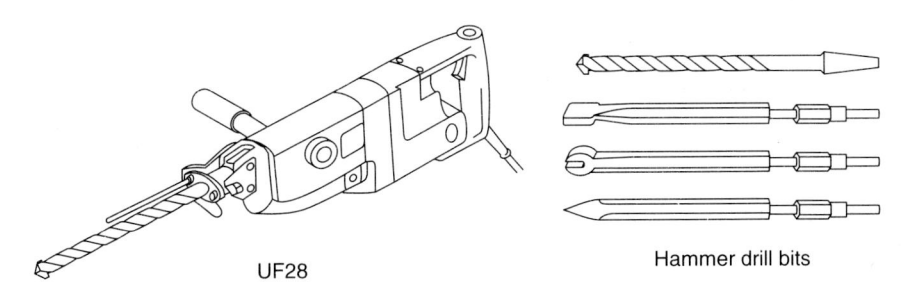

UF28

Hammer drill bits

Hammer drills have an 'A' taper shank and are suitable for use with rotary hammers for drilling into structural concrete, stone and engineering brick.

The drills have a toughened shank and are tipped with tungsten carbide. The combined performance of the rotary hammer and the hammer drill provides a greater number of holes more quickly than rotary impact methods.

Rawlcore Drills

The tubular construction of the Rawlcore drill means that only a narrow band of masonry is subjected to the tough carbide tipped cutting teeth, so that a fast rate of penetration is achieved. As a result, holes of up to 50mm (2") diameter are possible.

The Rawlcore drill is intended for use with rotary action machines only in boring large accurate holes in lightweight concrete, brickwork, tiles and masonry.

Method of Use

1. Commence drilling, using starter bit to ensure precise location of hole.

2. Remove starter bit after drilling for 3mm (1/8") and continue drilling, applying ample pressure and oscillating the drill regularly.

3. Clear spoil at intervals to prevent cutting head becoming packed. Do not stop the drill under load or whilst withdrawing it.

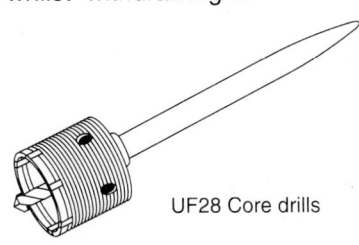

UF28 Core drills

Drilling Into Wood

For small holes, especially into softwood, a bradawl will be adequate. On hardwoods, spiral ratchet screwdrivers can be fitted with a drill bit, capable of making holes large enough to accept screw fixings for most types of electrical installations; switches, sockets, ceiling roses, saddles, brackets etc. A single speed hand drill with a range of drill bits from 2mm to 7mm (1/16" - 1/4") will be found adequate for many electrical installation purposes.

A two speed breast drill is useful for drilling holes larger than 10mm (3/8") into wood with the aid of twist drills.

A carpenter's brace and bit is invaluable when making screw fixings into wood. Screw auger bits are available which correspond in size with the usual range of screw gauges (diameters) in common use. Holes can be drilled through 6" thick wood by using the larger sizes of auger bits. Screwdriver bits and countersink bits can also be fitted. A ratchet brace is useful for drilling in awkward corners or in confined places where movement is restricted.

Single speed electric tools can also be used with twist drills or with round shanked power drilling bits designed for making deep holes into wood. Power tools will make drilling into awkward corners a lot easier and speed up the drilling process.

Drill bits range from bradawl and the grooved reciprocating bit to various types and sizes of auger bits, the twist drills and the power drills designed specially for boring into wood.

Drilling Into Wood

Holes for bolts or cables, etc.

1. Use a two speed breast drill and drills larger than 7mm (1/4") when drilling through wood for cable access etc.

Twist drills will make a clean cut hole through all woods, whether drilling across or into the grain.

2. The carpenter's brace and large diameter auger bits will drill large diameter holes cleanly only across the grain.

These drills are not really suitable for drilling into end grains.

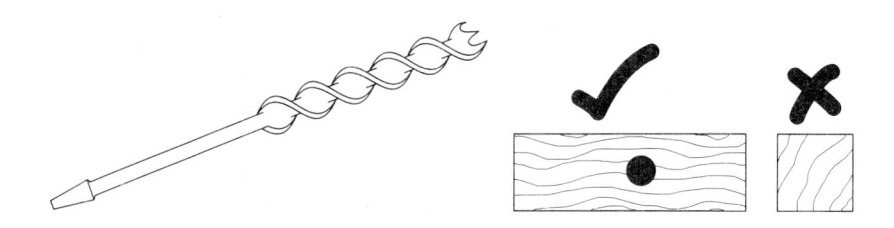

The centrally located screw tip literally screws itself into the wood with each clockwise turn of the brace, while the cutting edges cut around and lift layers of wood which are ejected by the spiralling action of the bit.

The power drill can be used with twist drills for all screw fixings into wood, metal or plastic materials; and with high speed wood drill bits to drill large diameter holes into wood.

4. When using power drills locate the tip of the drill over the centre point, before pressing the trigger switch to commence drilling. The tool must be held at right angles to the surface and the minimum pressure should be used to maintain accuracy.

Select the higher drilling speed for wood but never use the drill with trigger switch locked in the 'on' position.

Typical Applications

All the hand drilling methods described can be used with equal effect for making screw fixings into wood. Larger holes can be hand drilled but power drills are useful for repeated drilling operations of all kinds.

Safety

Note: Great care must be exercised when using power drills. Keep sleeves, ties and other clothing away from the revolving parts.

Use gloves to avoid injury from splinters.

Installing PVC Cables

Methods

The IEE Regulations Requirements for installing cables is that if it is not continuously supported, as is the case when pulled into conduit or trunking installations, then the cables should be supported by suitable means at appropriate intervals to prevent mechanical strain in the terminations of the conductors and the conductors themselves. For PVC cables this requirement can be met by use of plastic clips incorporating a masonry nail.

Guidance on spacing of clips for PVC insulated and sheath cables are specified in Table 4A of the IEE On-Site Guide to the 16th Edition IEE Wiring Regulations. Extract given below:

Cable Diameters mm	Support Spacings	
	Horizontal mm	Vertical mm
Up to 9	250	400
9 to 15	300	400
15 to 20	350	450
20 to 40	400	500

In locations such as under floor and behind partitions where cables are unlikely to be disturbed greater distances can be used. It will usually be found necessary to fix clips closer together especially on larger cables, if a neat appearance is to be achieved.

Surface Wiring

Where PVC cables are on the surface the cable should be run directly into the electrical accessory, ensuring that the outer sheathing of the cable is taken inside the accessory.

Concealed Wiring

If the cable is concealed, a flush box is usually provided at each control or outlet position.

Installing PVC Cable

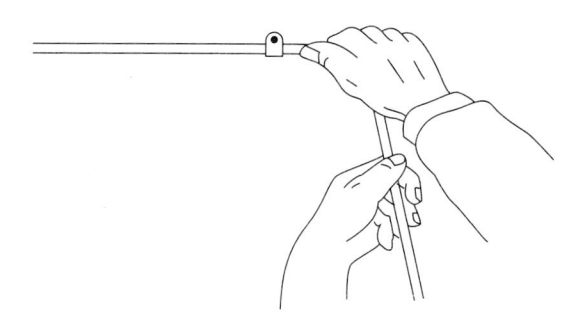

1. In order to ensure a neat appearance PVC cable should be pressed flat against the surface between cable clips.

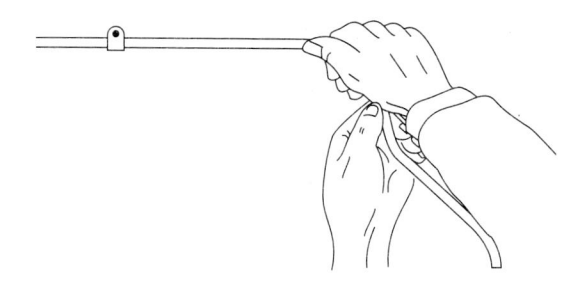

2. The cable should be formed by running the thumb against the surface of the cable, as illustrated.

Another method of forming the cable is to run the palm of the hand along the surface of the cable as illustrated.

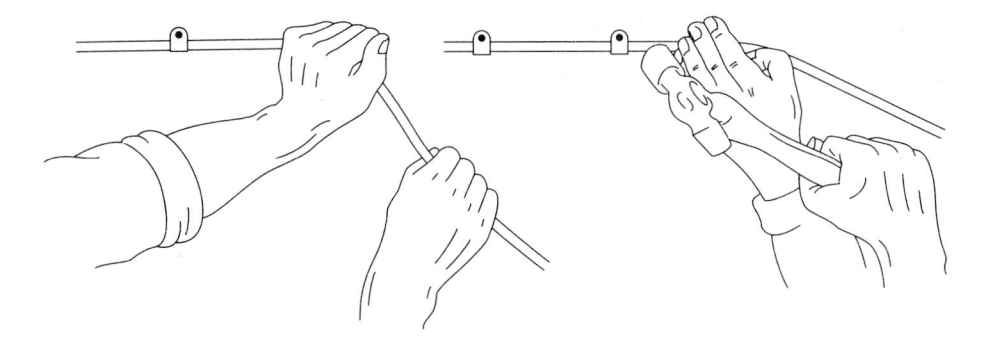

4. This sequence of forming the cable should be carried out after inserting the last cable clip and before fixing the next cable clip.

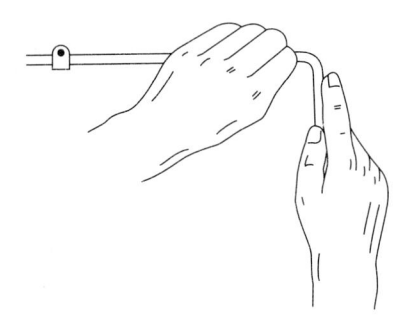

5. When a PVC cable is to be taken round a corner or changes direction, the bend should be formed using the thumb and fingers as shown.

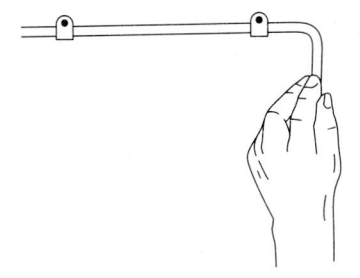

6. Care must be taken to ensure that the bend does not damage the cable or conductors. The cable must be supported at appropriate intervals so that it can support its own weight without damage.

Routing of PVC Cables

PVC cables are always vulnerable to mechanical damage, particularly where they cannot be seen easily.

The practice of covering cables with plaster is widespread and cases do occur of nails and other objects penetrating cables and causing damage. This gives rise to the risk of electric shock. In practice this risk seems to be small, nevertheless it is desirable to reduce the risk as much as possible.

There are two methods of reducing this risk:

Enclose the live conductors in earthed metal

Place the cables where they are less likely to be damaged.

IEE Regulation 522-06-07 requires the cable, when installed less than 50mm from the surface, to incorporate an earthed metal covering, or to be enclosed in conduit, trunking or the like. The enclosure must be substantial enough to fulfil the requirements of a protective conductor for the circuit in question, or by mechanical protection, sufficient to prevent damage to cables by nails or screws etc.

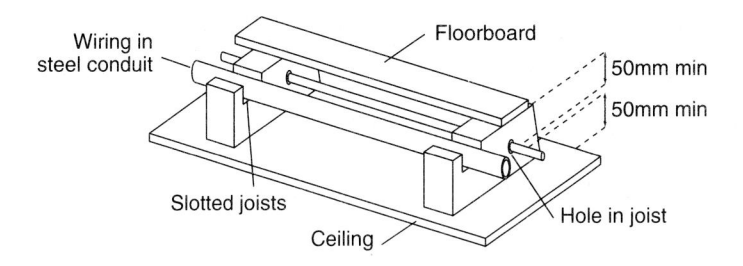

Even heavy gauge steel conduit does not give complete protection against mechanical damage and the cable may need to be replaced. Capping of metal or plastic is used to protect cables laid under plaster. This will protect the cables during the plastering operation, but gives very limited protection against nails and other objects driven into the plaster.

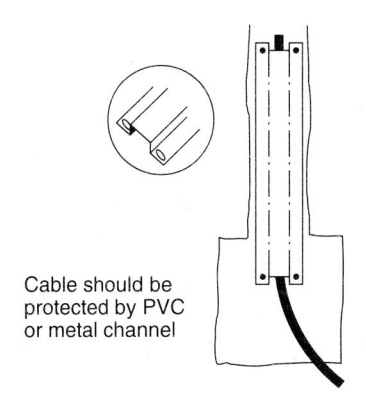

IEE Regulation (522-06-06) requires cables, not protected, to be placed where they are less likely to be damaged. The permitted zones are as follows:

(a) A strip 150mm wide along the top of the wall and alongside an adjacent wall or partition.

(b) A run either horizontally or vertically from the accessory to which it is connected.

In view of the practical problems in providing the earthed metal cover it is likely that cables will be installed mainly in the permitted zones.

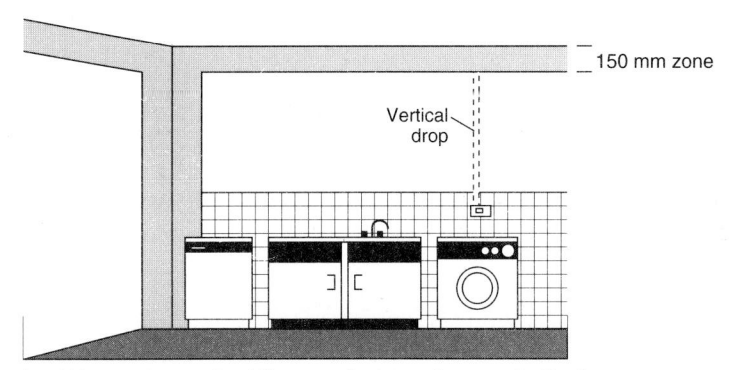

Cables should be run in permitted Zones or horizontally or vertically direct to accessory

Reducing the Spread of Fire Risk

IEE Regulation 527 requires steps to be taken when installing cables to reduce the risk of spread of fire.

Where a wiring system is required to pass through or penetrate material forming part of the construction of a building (e.g. cable, trunking or busbar trunking systems), areas external to the wiring system and where necessary internal areas, must be sealed to maintain the required fire resistance of the material.

Wiring systems with non flame propagating properties, having an internal cross section not exceeding 710mm_ need not be sealed internally. The sealing system used must meet the following requirements:

- be compatible with the wiring system concerned

- permit thermal movement of the wiring system without detriment to the sealing

- be removable without damage when additions to the wiring system are necessary

- be capable of resisting external influences to the same standards as the wiring system

During the installation of wiring systems, temporary sealing arrangements must be made. In addition, any existing sealing which is disturbed or removed in the course of alterations to an installation, must be reinstated as soon as possible.

It is essential that sealing arrangements are visually inspected during installation to verify that they conform to the manufacturer's instructions. Details of those parts of a building sealed and the methods used must be recorded.

Where cables pass through holes in metalwork, such as metal accessory boxes and luminaires, bushes or grommets must be fitted to prevent abrasion of the cables on any sharp edge.

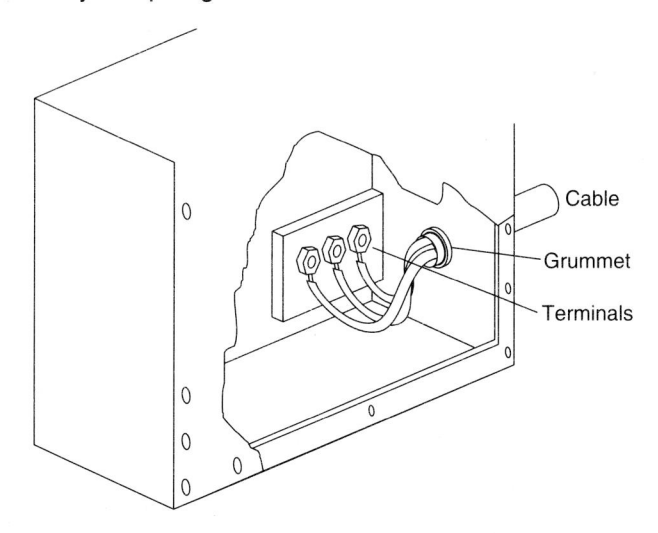

Keeping Cables Away from Other Service Installations

Care must be taken when installing PVC cables to ensure that they are not allowed to come into contact with gas pipes, water pipes and any non earthed metal work.

Fishing

When installing cables it is often necessary, when working in occupied premises to avoid moving furniture and lifting carpets unnecessarily. In these circumstances a technique called fishing for cables is used. This involves using galvanised wire known as the fishing wire and placing a hook on the end of the wire.

One piece of wire is placed at the point where the cable end is required and the other piece is pushed under the floor as illustrated. The wire is moved about until the other wire is touched and then slowly withdrawn until the hooks catch. The longest piece of wire is then pulled through with the short piece.

Fishing wire

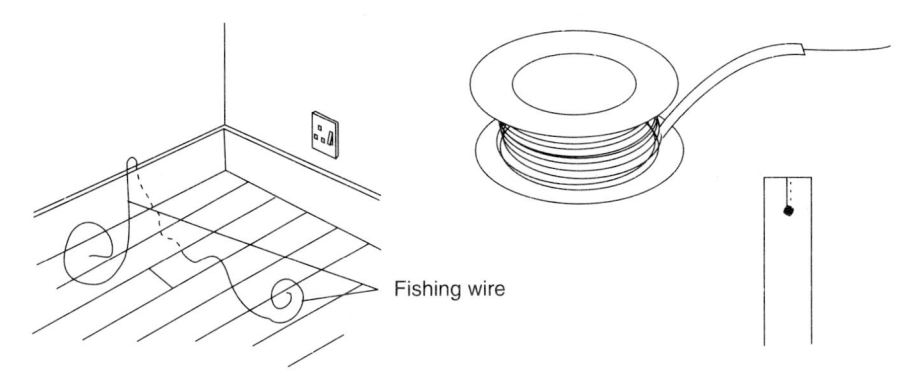

The cable is then fastened into the fishing wire to be installed using a bradawl and pliers and is pulled through as illustrated.

Identification Conductors

IEE Regulation 514 states that:

All cores of cables and conductors must be identified at the points of termination and preferably throughout their length to indicate their function. In some situations e.g. where a bathroom extractor fan with a timer circuit has been wired using 3-core cable with earth. With conductors coloured red, yellow and blue. The yellow and blue conductors will require marking at the points of termination in their functional colours; phase conductors (red) neutral conductors (black).

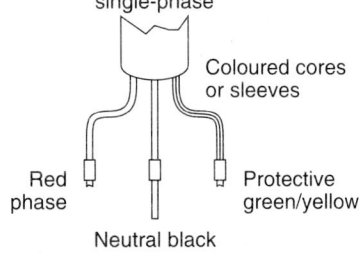

single-phase

Coloured cores or sleeves

Red phase
Protective green/yellow
Neutral black

Terminating Cables and Flexible Cords

The entry of a cable end into an accessory is known as a termination. In the case of a stranded conductor, the strands should be twisted together with pliers before terminating. Care must be taken not to damage the wires.

The IEE Regulations require that a cable termination of any kind should securely anchor all the wires of the conductor and not impose any appreciable mechanical stress on the terminal or socket. A termination under mechanical stress is liable to disconnection. When current is flowing a certain amount of heat is developed, and the consequent expansion and contraction may be sufficient to allow a conductor under stress, particularly one under tension, to be pulled out of the terminal or socket.

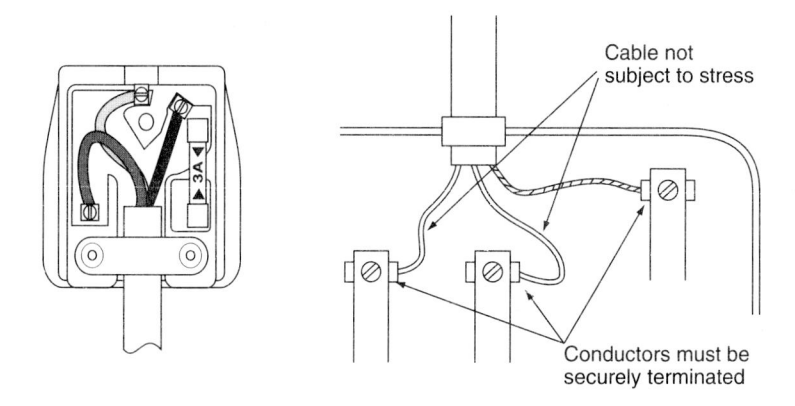

One or more strands, or wires, left out of the terminal or socket, will reduce the effective cross-sectional area of the conductor at that point. This may result in increased resistance and probably overheating.

Before a conductor can be terminated, the cable insulation should be removed.

Stripping Insulated Cables

The stripping of insulation is generally done with a knife. This should be sharp and held at a very acute angle to the cable so that the insulation is pared rather than cut. To cut at a near right angle would be very bad practice because, although it may give a neater finish, it would probably damage the conductor. A nicked conductor becomes so weak that after being bent a few times it will almost certainly break. Apart from this tendency to break the effective cross sectional area of the conductor will be reduced, causing increased resistance which may result in excessive heat. Wire strippers should be used if available, they make a cleaner, better job.

Method

1. Using side cutting pliers or a knife, slit the insulation from the end.

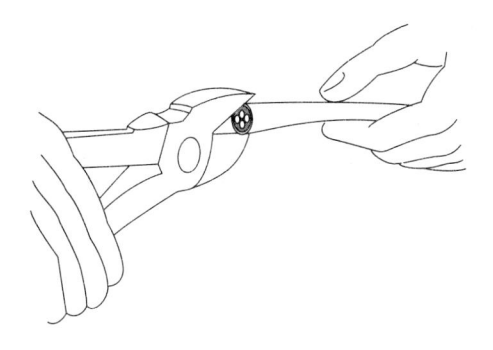

2. Then peel back the two halves of insulated sheath for a suitable distance.

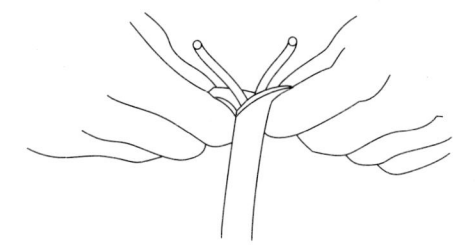

3. Then cut the sheath off, neatly.

4. Examine conductor insulation for damage.

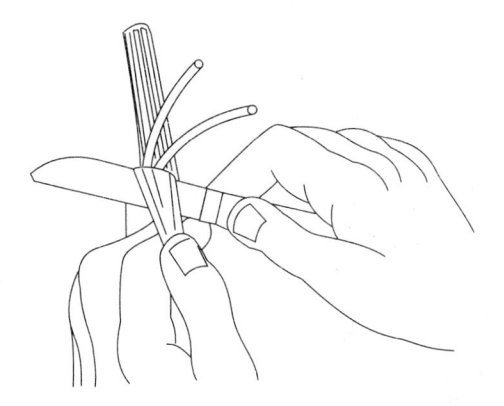

Wire Stripping Tools

1. Adjust screw to suit the diameter of the conductor. Push the end of wire into the tool so that the Vee slots close and cut through the insulation.

2. Remove severed insulated.

3. Examine the conductor for damage.

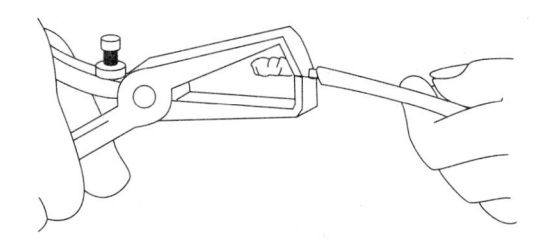

Type of Terminals

There is a wide variety of conductor terminations. Typical methods of securing conductors in accessories are pillar terminals, screwheads and nuts and washers.

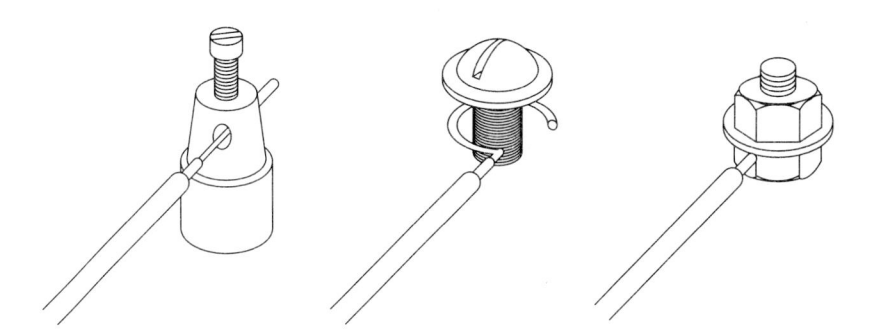

Pillar Terminals

A pillar terminal has a hole through its side into which the conductor is inserted and then secured by a set screw. If the conductor is small in relation to the hole it should be doubled back. When two or more conductors are to go into the same terminal, they should be twisted together. Care should be taken not to damage the conductor by excessive tightening.

Screwhead and Nut and Washer Terminals

When fastening conductors under screwheads or nuts, it is best to form the conductor end into an eye, using round nosed pliers. The eye should be slightly larger than the screw shank, but smaller than the outside diameter of the screwhead, nut or washers. The eye should be placed in such a way that rotation of the screwhead or nut tends to close the joint in the eye. If the eye is put the opposite way round, the motion of the screw or nut will tend to untwist the eye and will probably result in imperfect contact.

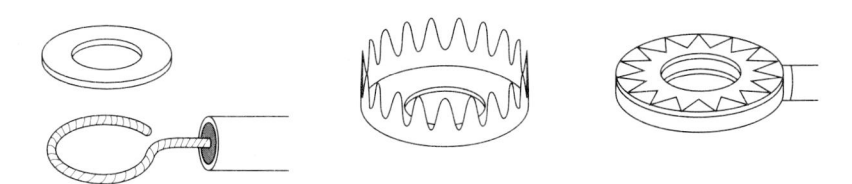

Claw Washers

In order to get a better connection claw washers can be used. Lay the looped conductor in the pressing. Place a plain washer on top of the loop and squeeze the metal points flat using the correct tool.

Strip Connectors

The conductors to be terminated are clamped by means of grub screws in connectors which are usually made of brass and mounted in a moulded insulated block. The conductors should be inserted as far as possible into the connector so that the pinch screw clamps the conductor. A good clean, tight termination is essential in order to avoid high resistance contacts resulting in overheating of the joint.

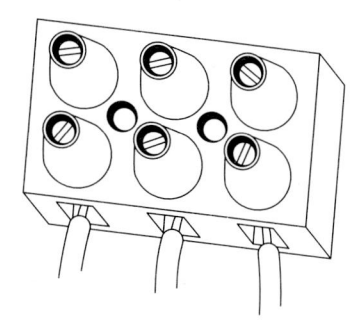

Terminating Cable Ends to Crimp Terminals

In order to terminate conductors effectively it is sometimes necessary to use crimp terminals. A situation where this type of terminal is required is the termination of bonding conductors to earth clamps and sink tops.

The terminals are usually made of tinned sheet copper with silver brazed seams.

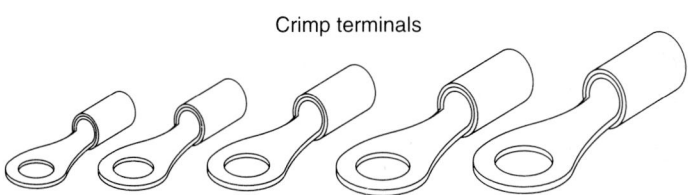

Crimp terminals

The crimping tool is made with special steel jaws which are adjustable in order that a range of cable and terminals can be crimped.

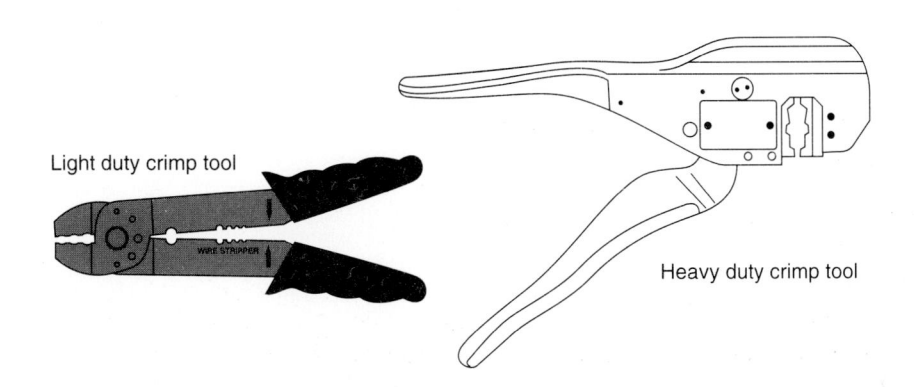

Light duty crimp tool

Heavy duty crimp tool

Method

1. Remove correct amount of cable insulation.

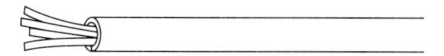

2. Place into terminal.

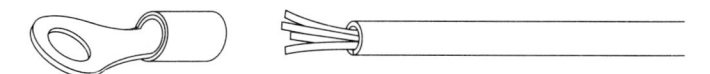

3. Crimp using crimping tool in accordance with manufacturers instruction.

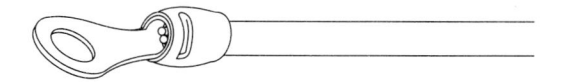

4. Check connection for soundness by holding cable firmly and giving terminal a sharp tug between thumb and forefinger.

Terminating a Plug to Flexible Cord

Tools Required

Insulated screwdrivers, sharp knife, wire cutters.

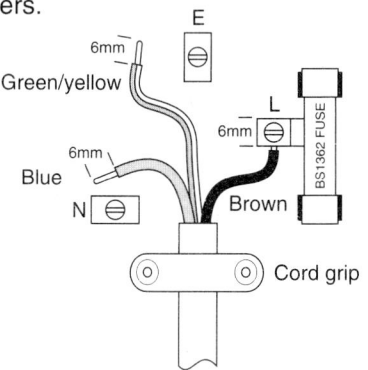

Unscrew or unlock the plug cover and remove it.

Gently prise out the cartridge fuse if necessary to reveal the terminal. Loosen the cord grip if necessary - plastic flanges grip the flexible cord in some plugs, in others plastic screws grip it.

Prepare the end of the flexible cord. For some plugs all the conductors have to be the same length. For others they have to be different lengths.

Bring flexible cord along plug to measure the required amount of outer covering to be removed.

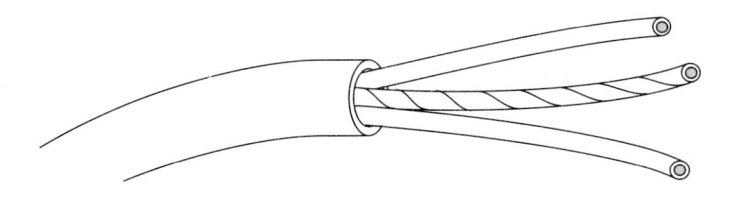

Cut conductors to correct length

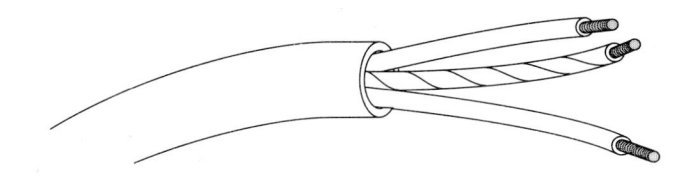

Remove sufficient insulation to expose conductor of correct length to ensure correct termination.

Fit the flexible cord into the plug pressing it between the flanges. If the plug is the type where the flex passes through a sleeve at the base of the cover before being clamped in place thread it through the sleeve.

Alternatively screw in place the bar that grips the flex. It is important to secure the outer sheath in the cord grip, not just the conductors. No unsheathed conductor must be visible outside the plug casing.

Connect the conductors to the correct terminals, securing them tightly in place. The brown conductor is connected to the terminal marked L, the blue conductor goes to the terminal marked N and the earth conductor (green and yellow) goes to the terminal marked E.

Wind each conductor up round the correct terminal and tighten the screw. Wind the conductor clockwise round the terminal. If you do not, it will be loosened as you tighten the screw and weaken the connection.

Alternatively if the terminal has a hole, push the conductor into the hole and tighten the screw.

Alternatively if the terminal is of the no screw type, push the conductor into the slot and swing over the clip to secure it.

The insulation must be close but not under the terminal clamping screws. Every strand or wire should be connected. Ensure that the terminal screws are tight.

Cord grip

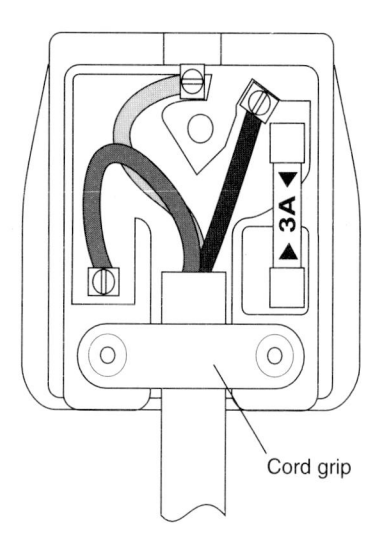

Replace the cartridge fuse if you had to remove it to reach the terminal. Make sure that it is of the correct current rating.

Screw the plug cover back in place. Alternatively, slide the cover in place and turn the plastic lock to secure it.

Visually inspect plug for correct assembly and tug flexible cord to ensure it is secured correctly.

Flooring

Lifting and Replacing Floorboards

Some wiring systems are run horizontally in the floor void and there may be occasions when floorboards will need to be lifted. This work must be undertaken with precision and care without damage to the floor.

Basic Preparations

1. Find out what it is you have to do. This will help you decide whether or not floorboards will need to be lifted, and their location.

2. Obtain the right tools and equipment for the job. These will include a bolster chisel, carpenters claw hammer, flooring saw or tenon saw, nail set punch and a short length of 3/4" - 1" thick batten or pipe.

3. Find the work area. In addition to knowing what it is you have to install, it is important to know exactly where you are required to install it. Check this with client or supervisor or consult working drawings.

4. Safety precautions:

First turn off the electricity supply at the main before commencing work.

Make sure that you know where the isolating points for gas and water are.

5. It is usual to try to trace back from the outlets to establish the direction of the cable run, so that where possible only a single floorboard need be lifted.

6. Examine the floorboards. Where there are no visible spaces between boards, it can be assumed that tongue and grooved floorboards have been used. (Some floors are not tongue and grooved).

7. If possible, find a section of board which has been butt jointed and work from that. Joints are usually obvious, the ends of the boards being nailed to the supporting joists.

8. Think the job through before you begin cutting and lifting the floor. Some time spent planning the job will usually save work and avoid unnecessary damage to the floor.

Tongue and Groove Flooring

Lifting Tongue and Groove Floorboards

It will be much easier to work if the room can be cleared of furnishings (i.e. carpets, lino, furniture, etc.) where this is not possible clear items to one side of the room. Take care not to damage them.
It is usual to try and trace back from the outlets to establish the direction in which the cables are running so that where possible only a single floorboard need be lifted.

Examine the floorboards and determine whether or not they are running in the same direction as the proposed installation. Check for visible spaces between floorboards using a knife or small screwdriver to confirm tongue and groove boards.

Lifting Method

1. Find a floorboard with a butt joint as near to the point of installation as possible.

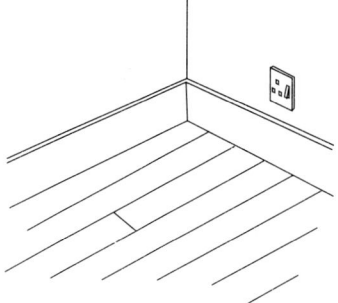

2. Start at the join and use a flooring saw to cut away the tongue on one side of the floorboard using the convex part of the saw to start the cut.

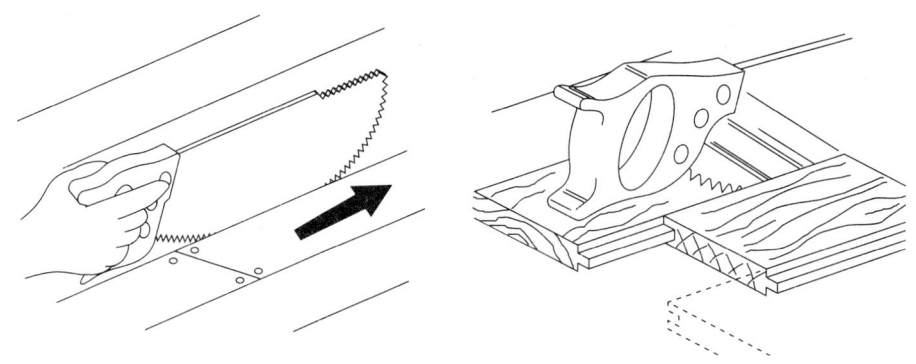

3. The tongue is situated in the centre of the floorboard and is usually 1/2" - 3/8" (6-10mm) thick.

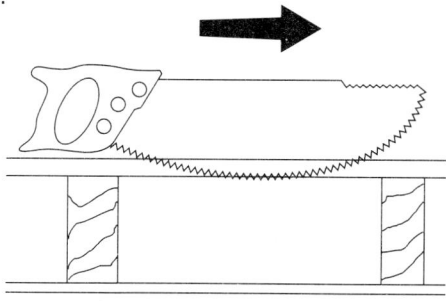

4. Continue sawing, holding the saw upright and level, allowing the saw to move forward with each stroke. Avoid cutting down through the board. Keep the saw moving forward in a straight line as the tongue is cut through.

5. Repeat the sawing process on the other side of the floorboard.

6. With the tongue cut through on both sides of the floorboard, position the point of the nail set on the head of the nails and hammer them down through the plank into the joists.

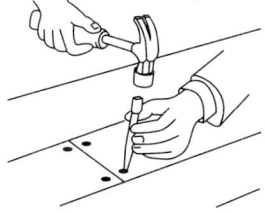

7. Insert a bolster chisel into the gap between boards about 4" (100mm) from the end and pull back to lever up the floorboards clear of the nail heads.

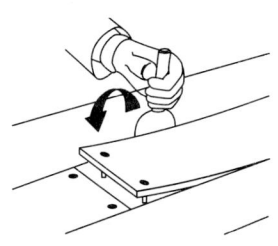

8. Move the bolster 12-18" (330-450mm) away from the join and lever the floorboards up sufficiently to allow a batten to be slid under the loosened floorboard to prevent it from falling back into place. (A short length of conduit will do!).

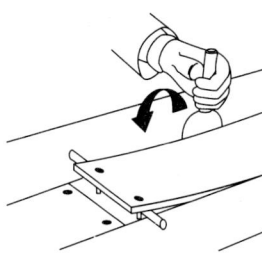

9. With the batten in this position continue cutting away more of the tongue on both sides of the board, sliding the rod or batten along towards the saw after each stage.

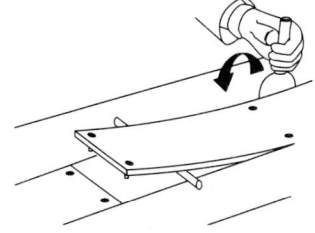

10. Use the bolster to prise the floorboard at points where it has been nailed to the joists. Position the chisel alongside the joist and push it into the gap as far as it will go before commencing to lever.

Push the batten along to help maintain leverage.

Lifting One-Length Floorboards

(Tongue and Groove Flooring)

One-length floorboards are boards which span the room in one unbroken length.

1. Starting in the centre of the room cut the tongues on both sides of the floorboard.
2. Punch nails through floorboard into joists.
3. Use a bolster chisel to prise up the floorboard as far as it will go.

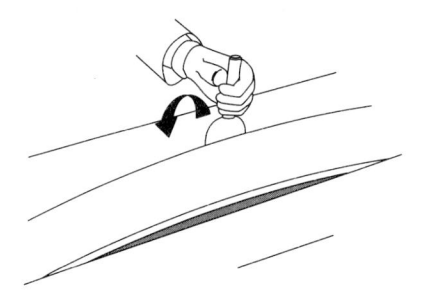

4. Now use a scrap piece of wood to increase the leverage and widen the gap to enable a batten to be slid underneath the partly raised floorboard.

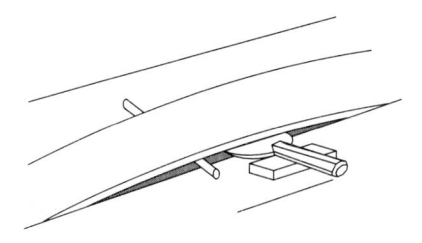

Patent floorboard lifters ("Delway" floorboard lifting and cutting systems) are available to do this operation. These mechanical aids will be particularly useful for large scale maintenance contracts.

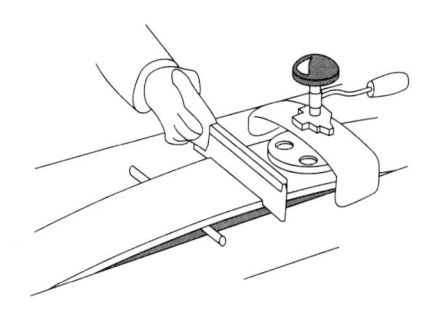

5. Shift the batten to rest over a joist and use the tenon saw to cut the floorboard squarely so that the cut lies centrally along the line of the joist. This will only be possible if the nails have been driven in or removed.

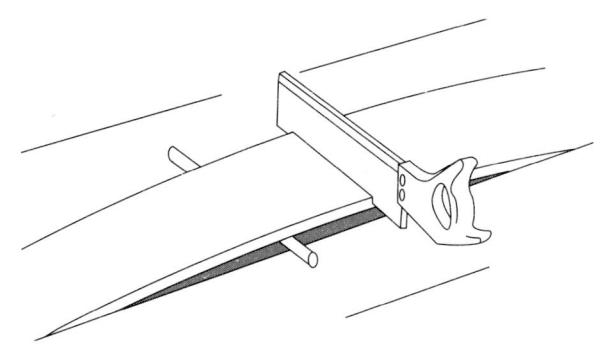

The alternative is to cut along the edge of the joist in which case a 25 x 50mm batten must be nailed to the joist to support the floorboard when it is replaced.

6. Slide the batten under the part to be lifted and use a bolster chisel to prise up the floorboard.

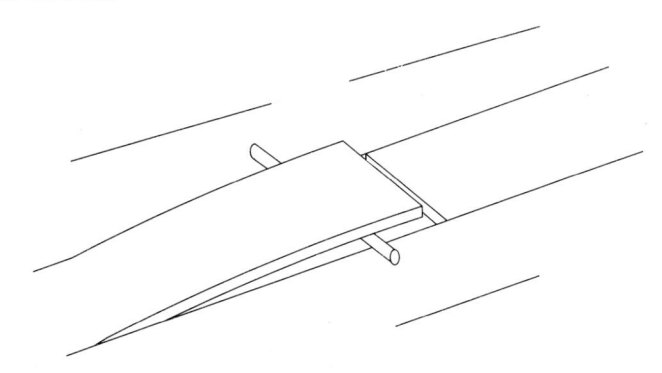

7. Move the batten along towards the wall at each step until the floorboard is completely free for lifting out.

Jig Saws

Jig saws provide an alternative method for cutting into floorboards. These are available in the form of attachments for power drills or purpose designed power tools.

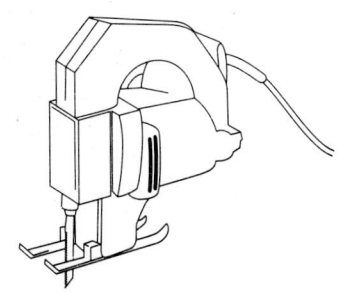

The important safety precaution is to ensure that the length of saw blade under the guide plate is equal to the thickness of the floorboard.

The saw blades can easily be ground down or broken off in a vice or cut with pliers, to the length required. Make certain that the saw blade is at the bottom of its cutting stroke when checking the blade length with the board.

Method of Use

1. First drill a small hole to accept the saw blade.

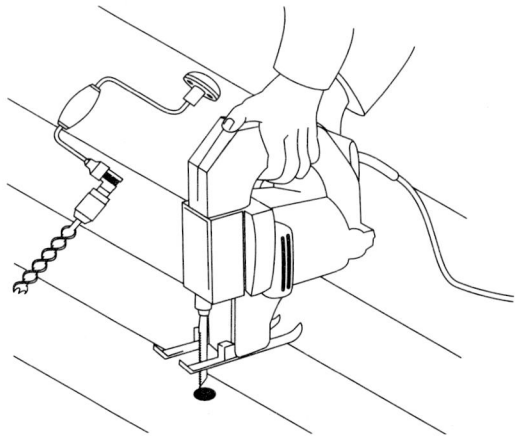

2. Alternatively holding the jig saw in this way, start the saw and allow it to cut its way into the floorboard until the guide plate is flush with the floorboard.

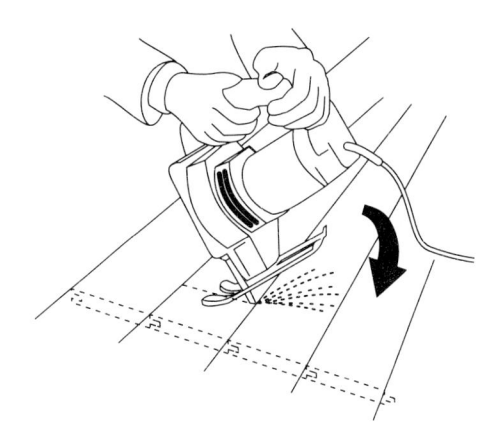

3. the jig saw is now correctly positioned to make the forward cut. While it is possible to make a reasonably straight cut freehand, it is advisable to use a piece of timber as a guide. When making this cut across the board, even though you have taken the precaution of shortening the blade, listen out for obstructions which are usually indicated by marked changes in the tone of the motor.

Timber guide

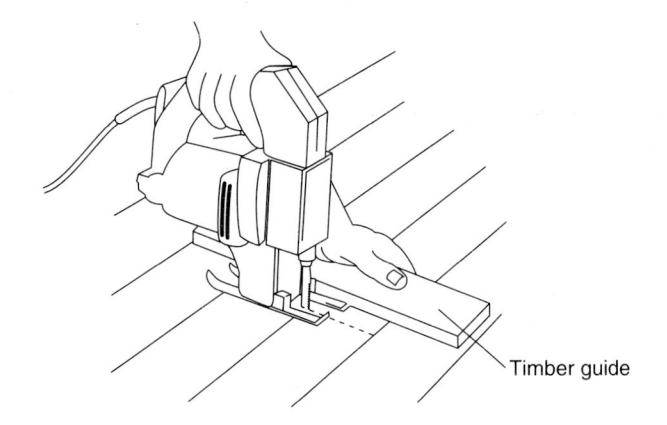

4. Cutting into the tongue of the floorboard is a lot easier than cutting across the board. A straight edge will ensure that the jig saw is cutting straight along the side of the floorboard.

5. Start the cut as already described at each corner before sawing forward using the straight edge as a guide until the floorboard is ready for lifting.

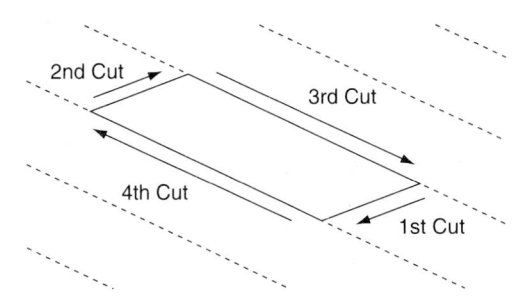

2nd Cut
3rd Cut
4th Cut
1st Cut

6. Use a bolster chisel to lever the floorboard up from where it has been nailed to the joists.

7. Remove all nails before laying the floorboard aside.

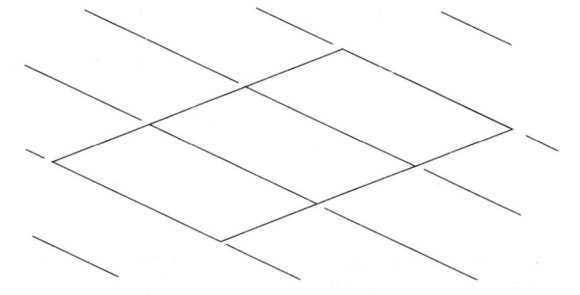

8. Where larger floor areas have to be lifted to permit access the same cutting methods are used to cut into the floorboards, except that it will only be necessary to cut through the tongues along two sides of the area chosen, and not both sides of each board.

Replacing Floorboards

Where possible floorboards should be replaced exactly in the same positions they occupied before they were lifted. Where single floorboards have been lifted, distinguishing features such as nail marks, saw cuts and discolouration etc. will help identify which end is to go where when replacing the floorboards.

Where more than one floorboard has been lifted it is worthwhile numbering the boards to show their relationship to each other and to other fixed boards. This will ensure they are returned to their former positions.

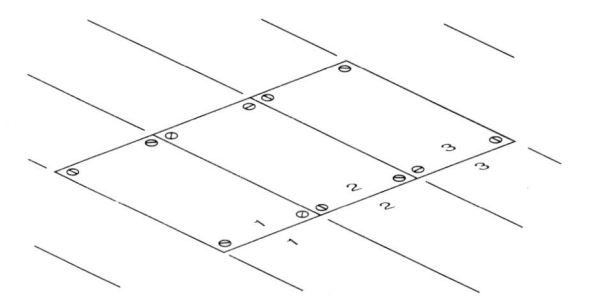

Additional Supports

Where more than one floorboard has been lifted, and especially where the boards have been left unsupported or too near the edge of the joist, it is advisable to nail battens along the joists to ensure the floor surface remains even and stable under foot. The 2" x 1" (50 x 25mm) supports are cut 4" (100mm) longer than the width of the trap and fixed with 2 1/2" (63mm) wire nails suitably spaced. Hold the batten flush with the floorboards and against the joist when completing the nailing operation.

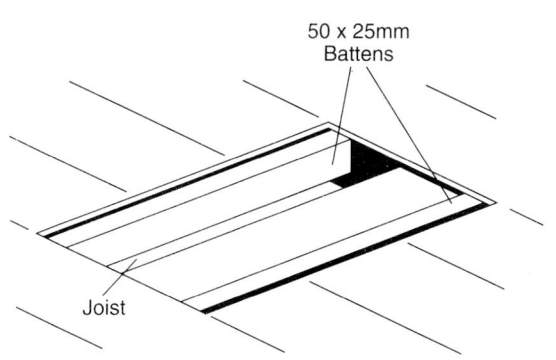

Securing the Floorboard

It is better to screw lifted floorboards back into place after completing any installation. Stagger the screw positions using two screws at both ends and one screw on intermediate joists making certain the screwheads are properly countersunk to obtain a 3/4" (19mm) depth of anchorage.

Chipboard Flooring

What is Chipboard?

Chipboard is made from wood particles bonded with a synthetic or organic agent. High density boards are often used on floors and these are usually 10 or 22mm thick. The boards are 2440-2745mm (8-9ft) long and come in widths of 400, 600, 1200 or 1220mm. It has a limited resistance to moisture and it is known to become permanently deformed when subjected to heavy loads over long periods.

Flooring grade chipboards are often used in place of softwood floorboards on standard joists which are 15" (380mm) between centres. They are butt jointed at the joists with the joints staggered. It is possible to find them grooved along both sides in which case a timber fillet (20 x 10mm) is used as the tongue.

Lifting Chipboard Flooring

Because these boards can be anything between 400 x 1220mm wide the room should be cleared of all furnishings to make the task of lifting easier.

Floorboards usually run at right angles to their supports and depending on what is to be installed and where it is to be positioned, a decision will need to be made as to which floorboard needs to be lifted. In practice instead of lifting up the whole floorboard, inspection traps are usually cut at convenient locations between joists.

Lifting Method

1. Use a knife or small screwdriver to test if there is a tongue between the boards. The tongues will offer solid resistance to the knife or screwdriver at a depth of between 6-10mm and will need to be cut through.

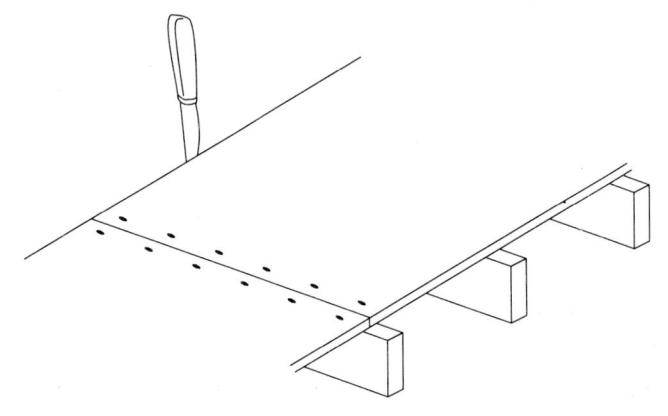

2. Using the convex part of a flooring saw start, cutting into the tongue at the mid-point between any two joists. The forward part of the saw can later be used to complete the cut at the corners.

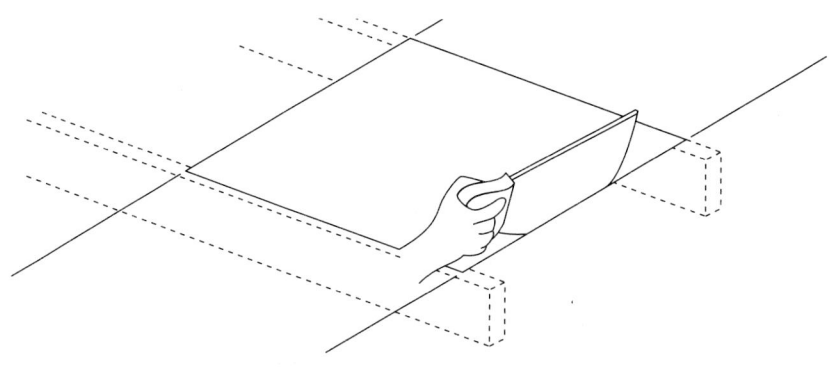

3. The tongue is situated in the centre of the floorboard and is usually 1/4" - 3/5" (6-10mm) thick.

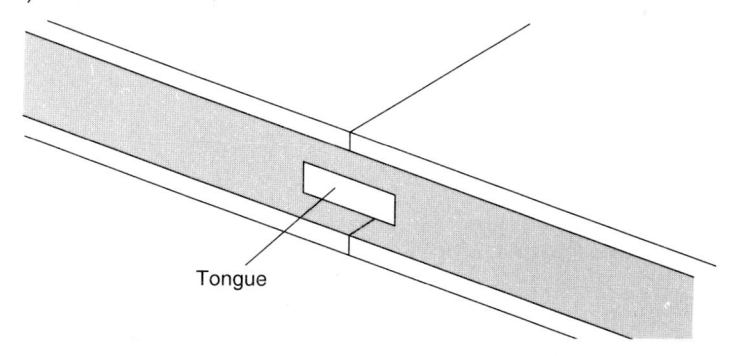

4. Saw cuts will often extend beyond the corners particularly where these happen to lie over the joists.

5. Avoid cutting down through the floorboard. Keep the saw moving forward in a straight line as the board is cut through.

6. After completing the first cut (A) it is necessary to mark out the other cuts using a try square. Cut B should run along the edge of the joist and be at least as long as cut A.

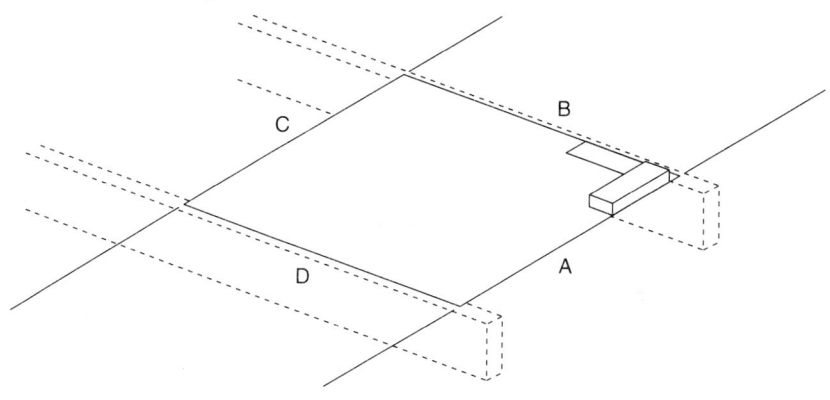

7. Where 400mm or 600mm wide chipboards have been used, cut B will extend to the full width of these boards.

400 or 600 wide chipboards

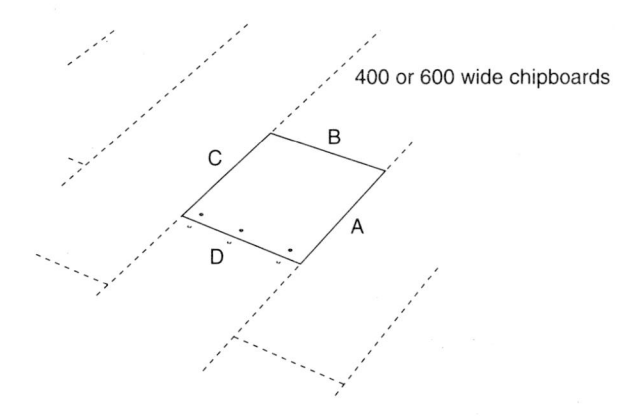

8. Start the cut in the centre using the convex part of the flooring saw until the board is sufficiently cut through. The saw can now be used at an angle to complete cutting to the corners of the trap.

Remember when sawing blind that there may be hidden hazards - cables, pipes etc. under the floorboard.

9. Use the same technique, starting at the middle and working outwards to the corners.

10. Where there is a join in the floorboards located over a joist, raise the flap which has now been cut through on three sides and prise the nails clear of the joist, using the bolster chisel to lever with.

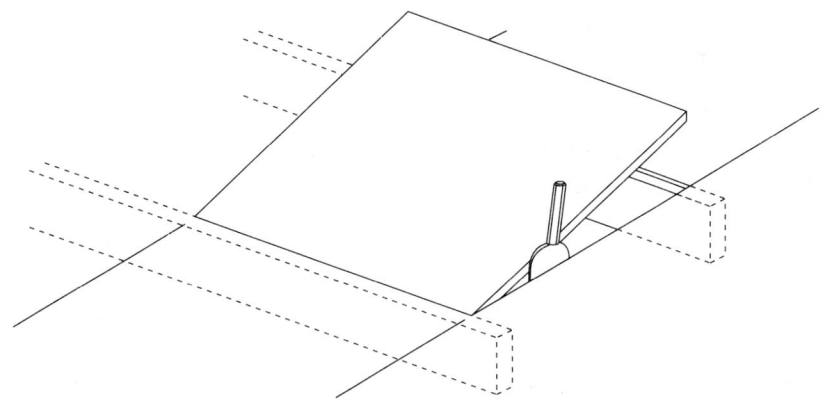

11. Where the trap is being cut in the centre of the board, it will be necessary to make a fourth cut along the edge of a joist.

Replacing Traps

Whenever traps are cut in chipboard floors it is always necessary to put in some timber framework to support the piece which has been cut out. As a general rule, support for most traps should be provided on all sides. The easiest way of providing this is explained below.

Butt Joined Frames

1. Cut timbers 6" (150mm) longer than the size of the opening.

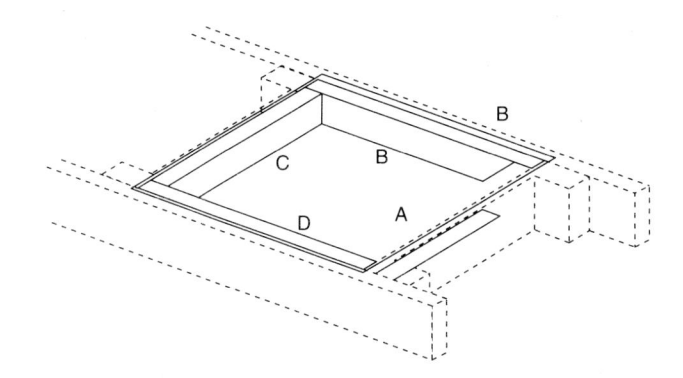

2. As a first step drive two 3 1/2" oval nails partly through the timbers ready for nailing to the joists. Locate each centrally and level with the top of the joist and complete the nailing operation.

3. With both timbers fixed in position, measure the distance between these and cut timbers to fit these gaps.

4. Drive a single 3 1/2" oval nail at an angle into each end. The object is to nail these pieces to the timbers which have been securely fixed to the joist.

5. Hold the pieces in squarely making sure that a 1" lip is left for the trap cover to rest on.

Complete nailing operations at both ends of each timber.

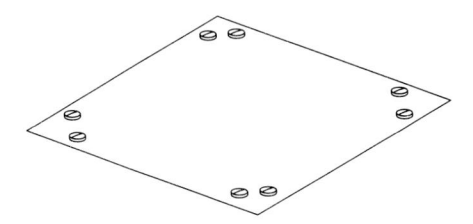

6. Nail down the trap opening and screw down the trap cover as previously described.

Alternative Method

Where the trap opening is reasonably square it will be possible to make up the frame by nailing through the longest timbers into the ends of the shorter timbers and simply sliding the frame into position for nailing to the joists.

Chasing Work (Raggling)

Concealing Cables

It is often necessary to conceal cables in walls, floors and ceilings. Two methods of concealment are generally used:

wiring systems installed in floors, walls and ceiling spaces

wiring systems embedded in floors, walls and ceilings

Cables Installed in Floors, Walls and Ceilings

Wiring systems should only be concealed in floor or ceiling voids, or in internal wall spaces; not in external cavities, for the following reasons:

PVC insulated cable installed in an external wall cavity may be adversely affected by the introduction of cavity wall insulation and this may result in the cable overheating.

The cable may also 'bridge' the air gap and cause moisture to be transmitted from exterior to under wall surface, via the cable.

Embedded Wiring Systems

Metal or plastic conduits and PVC cables are generally fixed to the fabric of the building before the walls are plastered or before the concrete is poured. Conduit and terminal boxes should be sealed by wood plugs or paper to prevent the entry of cement or plaster.

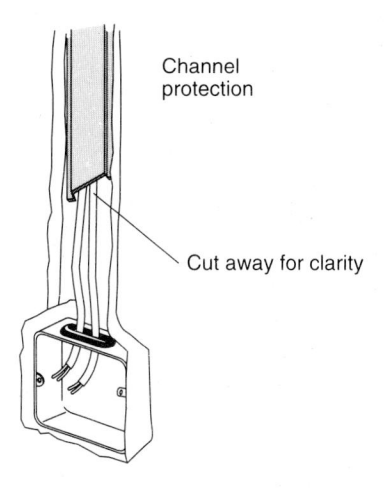

Channel protection

Cut away for clarity

Wiring systems can also be concealed in a chase cut in the plaster or brickwork. The chase must be sufficiently deep to allow a minimum of 5mm plaster skim. Anything less than this would permit rust from conduit, channel or fixing screws to penetrate to the surface.

Embedded PVC Cables

In instances where wiring is installed during the course of the construction of the building, oval conduits or metal or plastic channelling should be used to protect cables from damage.

Where walls are already plastered a chase must be cut in the plaster and brickwork. Sheathed cables should be securely fixed and protected by channelling. Plaster mixes such as 'Poly-filla' etc. are used to make good.

Marking Out Chases

When chasing into building surfaces the aim should be to minimise disturbance to the surface. This is best achieved by marking off parallel lines just wide enough for the job in hand.

Safety Precautions

Safety goggles must be used whenever a chase is cut to prevent flying particles causing injury to the eyes. Suitable gloves should also be worn.

A plastic sheet should be placed on the floor to catch most of the debris and to make the job of clearing up easier.

Chase Work Using Hand Tools

Cutting a Chase in Plaster and Brick

1. Mark and chase by drawing parallel lines to show the width of chase to be cut and the size of terminal box to be fitted.

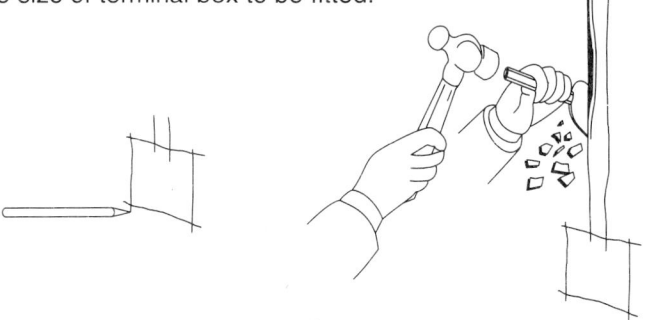

2. Lay a plastic sheet on the floor to catch the bulk of the debris.

3. Put on safety goggles and use a pair of gloves.

4. Use a wide flat chisel and light taps from the hammer to outline the chase and terminal box.

Angle the chisel slightly inwards.

Channel protection

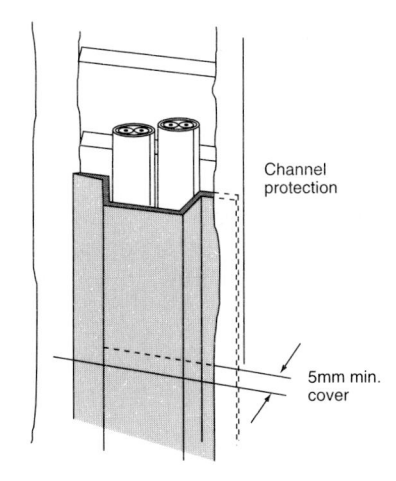

5mm min. cover

5. Chisel through the plaster down to the brickwork and lever off the plaster from the area being chased to expose the brickwork underneath.

6. Check the thickness of the cables and other mechanical protection (channelling, clips, over conduit, etc.) to be laid against the plaster.

The level of the cables or channelling should be at least 5mm below the surface of the plaster for proper concealment.

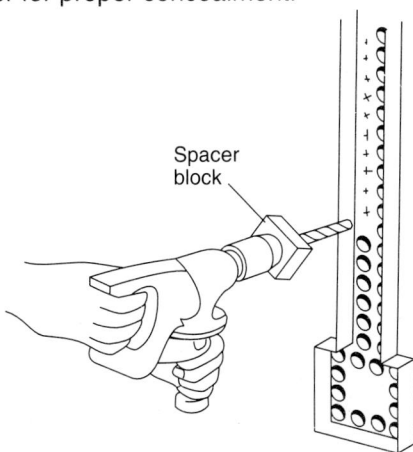

Spacer block

7. Where chasing into the brickwork is necessary, use the two speed electric drill at the slowest speed and a masonry drill (No. 10 or 12). Drill holes right around the area cleared of plaster.

To obtain the correct depth of drilled hole use a block of wood of the required thickness as a gauge. Simply drill centrally through the block of wood taking care to hold it firmly when drilling. In this way holes of the correct depth can be drilled each time. If a proper gauge can be fitted to the drill, use it in preference to a block of wood.

Holes should not be more than 5mm apart around the sides of the chase.

8. A third vertical row of holes down the centre of the chase may also be necessary to facilitate chiselling away the intervening brickwork.

The area to be recessed for the terminal box will need to be honeycombed with drill holes, using a spacer block of reduced thickness to obtain holes of the correct depth for the box to be fitted.

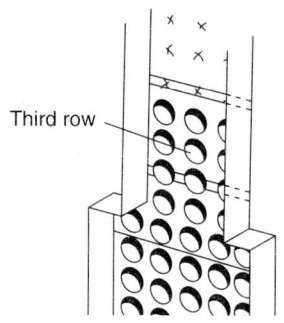

Third row

9. Using a sharp cold chisel and short sharp strokes from the hammer to even up the bottom and sides of the chase down to the required depth.

Position yourself to one side of the chase when chiselling to keep out of the way of the flying particles and dust, but make certain that you can clearly see what you are doing.

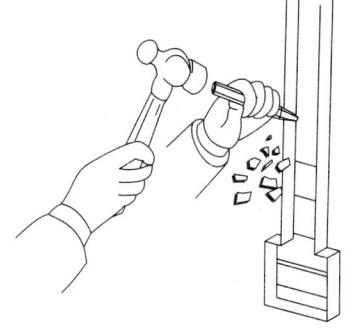

10. Chisel out the recess for the terminal block again using the cold chisel.

Complete the recess by chamfering away the vertical chase to allow for a smooth entry of the cables into the terminal box.

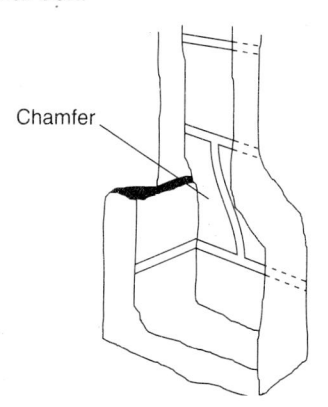

Chamfer

11. Check that the lip of the box is 5mm below the level of the plaster surface. Position the terminal box so that it is level and plumb. Use a small spirit level to check.

Mark out the drill holes, drill, plug and screw the terminal box into position.

12. Draw in the sheathed cables through the rubber grommet and fix channelling over the cables using galvanised or non-ferrous nails.

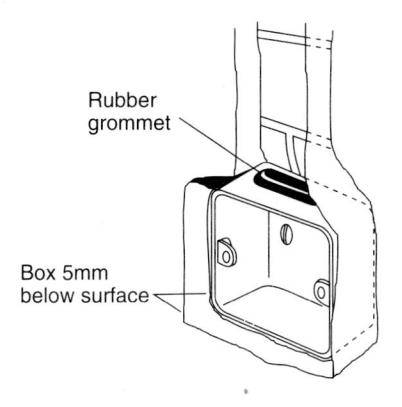

Rubber grommet

Box 5mm below surface

13. Mix up sufficient plaster and water to fill one chase at a time.

Do the mixing on a board adding the water (1 part by volume to 6-7 parts of plaster) slowly to a hollow made in the centre of the heap of plaster. The mix should be quite dry but smooth (not lumpy).

14. Push some crumpled newspaper into the switch box to keep it clear of plaster. Wet the chase to prevent plaster drying out too quickly and cracking.

Place some of the plaster on a smaller board hold it against the wall and starting at the bottom use the trowel to push small quantities of plaster at a time into the chase to provide a rough rendering coat to fill 2/3 of the space.

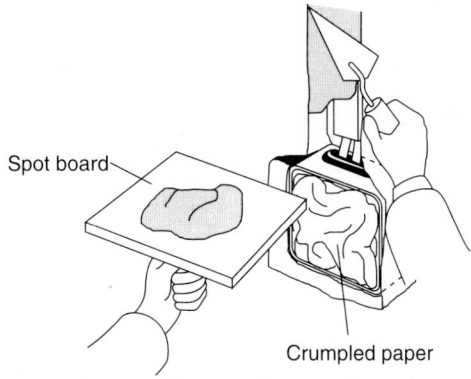

Spot board

Crumpled paper

15. Next use the tip of the trowel to fill and finish off the plastering around the terminal box and complete filling in the chase with plaster. For deep chases (in excess of 25mm) it is advisable to wait 24 hours before completing the filling operation to allow the rendering coat to set reasonably solid.

16. Sprinkle the plaster with minute quantities of water (using a brush dipped

in clean water) and smooth out the surface of the plaster.

Remove paper from accessory box and connect the switch or socket outlet.

Bathroom and Showers

In rooms containing a fixed bath or shower where permanent and reliable metal pipework has been installed, supplementary bonding conductors shall be installed as illustrated below, to reduce to a minimum the risk of electric shock in circumstances when the body resistance is likely to be low (601-04-02). The most common method of making a bonding connection to pipework is by using BS 951 earth clips.

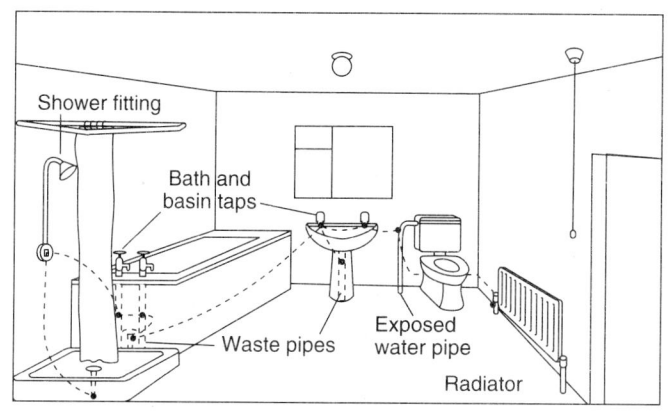

In other cases it is necessary to bond hot, cold and heating system pipework at one point only e.g. airing cupboard. This requirement does not apply to equipment supplied from a SELV circuit.

All circuits and equipment to be protected by protective devices which operate in 0.4 seconds.

No electrical equipment shall be installed in a room containing a bath or shower which is within reach of a person using the bath or shower unless it is a shaver socket to BS 3535 (fitted with an isolating transformer) or electrical equipment and control circuits operating up to a maximum of 12 volts supplied from a remote safety isolating transformer.

For example, the installation for a whirlpool bath. When the electrical equipment is installed below the bath e.g. pumps, the space must only be accessible by the use of a tool e.g. screwdriver.

Switches for lighting and fixed equipment such as showers should be of the insulating cord operated type. Positioned such to allow the cord to be within reach of a person using a bath or shower if necessary but not the switch itself.
All luminaires (light fittings) within horizontal distance of two and a half metres of a bath or shower should be totally enclosed.

Any equipment installed in a bathroom or room containing a shower or sauna should be suitable for that environment i.e. splashing of water and high humidity levels.

Surface wiring systems used in bathrooms must not have exposed metal covering e.g. metal conduit or exposed earthing or bonding conductors e.g. without insulation.

Shower Cubicles Not in a Bathroom

When shower cubicles are installed, for example in a bedroom, no socket outlets should be within 2.5 metres of the shower cubicle doorway. Those socket outlets that are installed in the same room as the shower cubicle should be protected by an RCD.

Any bayonet - type lampholders in that room should be fitted with home office skirts.

Identification Notices

Switchgear Control Gear and Protective Devices

Switchgear and control gear in an installation should be labelled to indicate its uses. Where the operation of switchgear or control gear cannot be seen by the operator, an indicator light or other signal should be installed. British Standard 4099 deals with the colours which should be used for indicator lights, push buttons, annunciators and digital readouts.

All protective devices in an installation should be arranged and identified so that their respective circuits may be easily recognised.

Diagrams

Diagrams and charts must be provided for every electrical installation indicating:

(a) the type of circuits

(b) the number of points installed

(c) the number and size of conductors

(d) the type of wiring system

(e) the location and types of protective devices and isolation and switching devices

(f) details of the characteristics of the protective devices for automatic disconnection, the earthing arrangements for the installation and the impedances of the circuit concerned.

(g) circuit or equipment vulnerable to a typical test e.g. equipment with electronic components such as central heating programmers.

The purpose of providing diagrams, charts and tables for an installation is so that it can be inspected and tested in accordance with Chapter 71 of the IEE Regulations and to provide any new owner of the premises (should the property change hands) with the fullest possible information concerning the electrical installation.

It is essential that diagrams, charts and tables are kept up to date.
Typical charts and diagrams for a small installation are illustrated below.

Schedule of installation at ....

Type of circuit	Points served	Phase Conductor	Protective Conductor	Protective devices	Type of wiring
Lighting	10 downstairs	$1mm^2$	$1mm^2$	5 Amp Type 2 MCB	PVC/PVC
Lighting	8 upstairs	$1mm^2$	$1mm^2$	5 Amp Type 2 MCB	PVC/PVC
Immersion heater	Landing	$2.5mm^2$	$1.5mm^2$	15 Amp Type 2 MCB	PVC/PVC
Ring	10 downstairs	$2.5mm^2$	$1.5mm^2$	30 Amp Type 2 MCB	PVC/PVC
Ring	8 upstairs	$2.5mm^2$	$1.5mm^2$	30 Amp Type 2 MCB	PVC/PVC
Shower	Bathroom	$6mm^2$	$2.5mm^2$	30 Amp Type 2 MCB	PVC/PVC

Warning Notices

A warning notice stating the maximum voltage present should be fixed to every item of equipment (or enclosure) which contains circuits operating at voltages in excess of 250 volts and where the presence of such a voltage would not normally be expected.

Where accessories, control gear or switchgear are wired on different phases of a 3-phase supply, but can be reached simultaneously, a notice must be placed in a position where anyone removing an accessory, or gaining access to the terminals of control gear, switchgear etc. is warned of the maximum voltage.

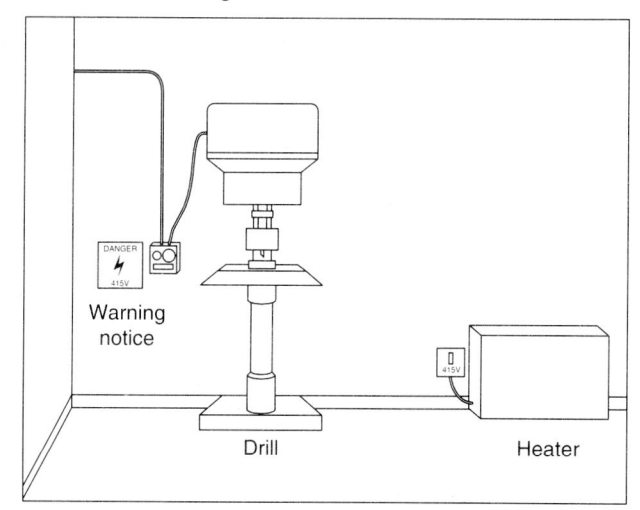

Inspection and Testing (514-12)

Upon completion of an electrical installation the electrical contractor should fix in a prominent position on or near the main distribution board of the installation, a label with details of the date of the last inspection and the recommended date of the next inspection.

The notice must be inscribed with characters (not smaller than 11 point), as illustrated below.

'IMPORTANT'

This installation should be periodically inspected and tested, and a report on its condition obtained, as prescribed in the Regulations for Electrical Installations issued by the Institution of Electrical Engineers.
Date of last inspection........................
Recommended date of next inspection ....

Residual Current Device - Notices

When an installation incorporates a residual current device, a notice must be fixed in a prominent position, at or near the main distribution board. It should be printed in indelible characters, not less than 11 point in size and should read as follows:

> This installation, or part of it, is protected by a device which automatically switches off the supply if an earth fault develops. Test quarterly by pressing the button marked 'T' or 'Test'. The device should switch off the supply and should then be switched on to restore the supply. If the device does not switch off the supply when the button is pressed, seek expert advice.

Earthing and Bonding

A warning notice (as illustrated) must be fitted in a visible position near to the point of connection of an earthing conductor to an earth electrode, or a bonding conductor to an extraneous conductive part.

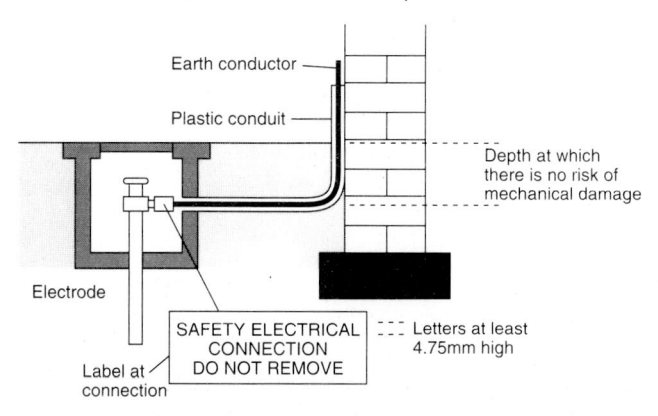

Earth conductor
Plastic conduit
Depth at which there is no risk of mechanical damage
Electrode
Label at connection
SAFETY ELECTRICAL CONNECTION DO NOT REMOVE
Letters at least 4.75mm high

Electrically Operated Controls

Introduction

Gas fired central heating systems commonly have electrically operated controls, so that the heating system can be made to operate to meet the consumer's needs; to give safety and economy in operation and to assist in maintaining the desired comfort conditions during specific hours of the day or night.

Heating systems can be controlled by any or all of the following methods:

by water temperature in the system

by the temperature in the appliance

by reference to the air temperature within the premises

by reference to the outside temperature

on a time basis

These controls can be divided into two categories: switches and components. Switches are usually operated by some mechanism; for example, a heat sensitive element (thermostat) or controlled by a clock. Components include motors, pumps, gas valves, ignition coils etc.

Operating Mechanisms

Various types of switches are used in central heating systems, but they should all satisfy the following requirements:

In the off or open position - no continuity
In the on or closed position - continuous circuit

Switches can be of various types: single-pole; double-pole; change over or rotary switches which can be operated manually, mechanically or by electro-mechanical means, or be heat operated.

Thermostats

An electric thermostat is a device which operates a switch in response to change of temperature. It consists of two main parts:

(a) a temperature sensitive element
(b) a switch

The temperature sensitive element is usually a bi-metallic strip; or it may be a differential expansion mechanism or a phial and bellows mechanism.

Bi-metal Strip

These are usually found in room thermostats and for controlling the temperature of air heaters. The sensitive element consists of a junction of two dissimilar metals which bend with change of temperature. This is used to operate a switch. In order to get a large movement for a relatively small temperature change a long bi-metallic strip is sometimes wound into a spiral or helix.

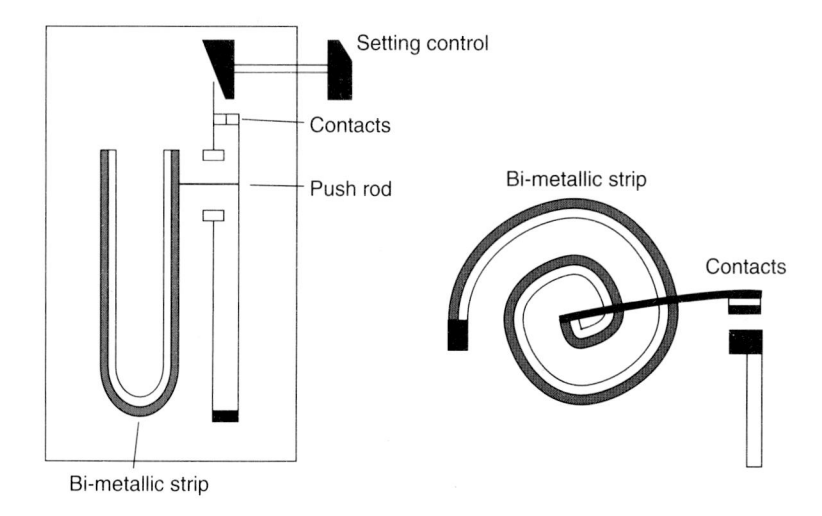

Phial and Bellows

This type of mechanism is least common in room thermostats. The system may be partly filled with a volatile liquid, in which case the distension of the bellows will depend on the vapour pressure of the liquid; or it may be completely filled so that the movement of the bellows depends on the coefficient of expansion of the liquid. Whichever type of actuator is used, its movement is applied to operate the switch.

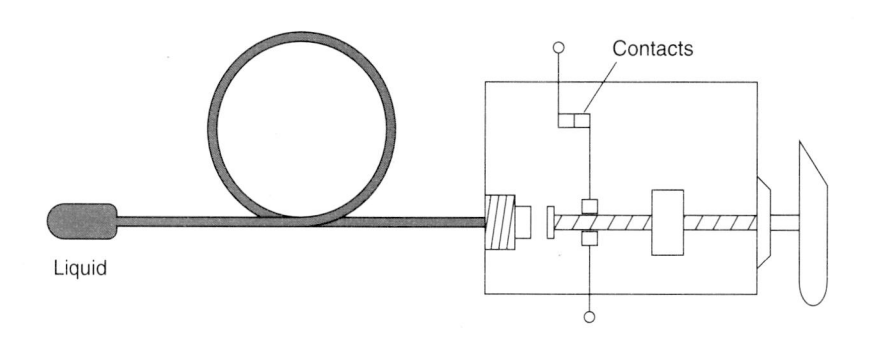

Differential Expansion

This consists of an (invar) rod within a brass tube. The two metals expand at different rates, (differential expansion). This movement is accentuated by means of a lever and is so arranged as to operate a pair of pre-set electrical contacts.

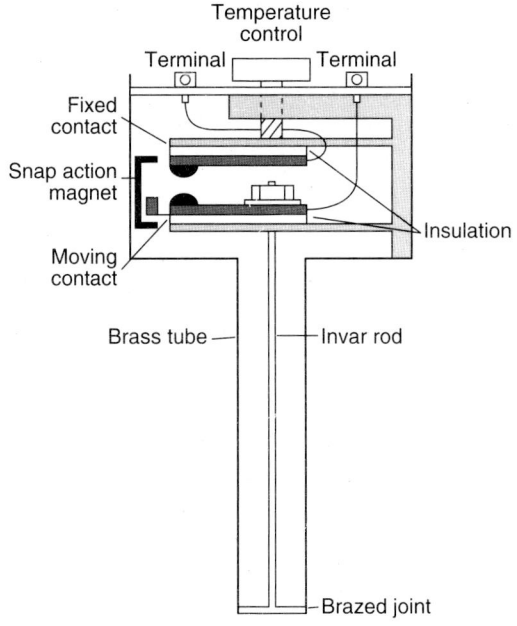

Thermocouple

The thermocouple is a device for detecting relatively large changes in temperature at specific locations. Its main use in gas installations is to shut off the gas supply to a pilot light if the flame is extinguished.

If two dissimilar metals (frequently iron and eureka) are joined together at one point and the junction exposed to a heat source the different levels in molecular activity will cause an electric current to flow which can be detected and used to operate a valve, cutting off the supply of gas. A diagram illustrating this principle is shown below:

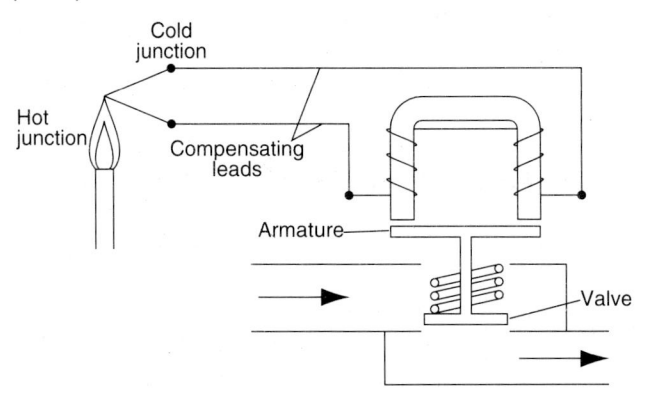

In use the thermocouple has proved to be extremely reliable. The common faults and causes of failure include an inadequate pilot flame or one exposed to draughts; the accumulation of dirt and soot or physical damage due to a sharp blow or incompetent installation. The main advantage of the thermocouple is that it is a "fail-safe" device, that is to say that in the event of failure the gas supply will automatically be shut off. In the event of a fault the unit should be replaced.

The Switch

The switch may be direct-acting contacts, or magnetic assisted, direct-acting contacts, a mercury switch or a micro-switch.

With direct-acting contacts, one end of the temperature sensitive element is fixed to the thermostat body and the other carries a contact which can engage with a fixed contact. This type of mechanism is rarely used, as a slow change of temperature causes slow opening and closing of the contacts, which may result in arcing of the electric current between the contacts.

This problem of slow and imprecise opening of the contacts is overcome in a number of ways. In a magnetic assisted, direct-acting switch a contact is mounted on a flat spring which can be deflected by the sensitive element. (If the sensitive element is a flat bi-metallic strip the moving contact is fixed directly to it.) Fixed to the spring is a soft iron armature which is attracted by a small permanent magnet fixed to the body.

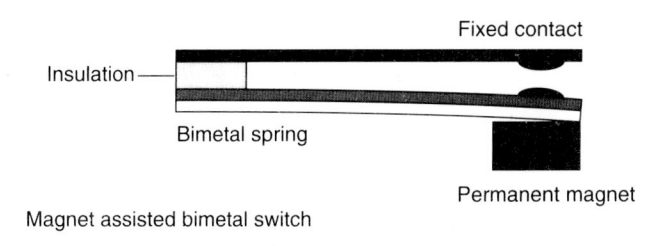

Magnet assisted bimetal switch

Magnetic attraction is inversely proportional to the square of the distance. When the switch is fully open there is little attraction but as the contacts approach, the pull increases rapidly until it overcomes the spring and snaps the contacts quickly together.
When the thermostat operation reverses, a considerable pull must be applied to separate the armature from the magnet, after which the contacts fly open.

Micro-Switch

A micro-switch is a switch which will operate from a very small movement. Micro-switches are to be found in thermostats for use in warm air systems for air temperature control and fan control.

Mercury Switch

A mercury switch as used in thermostats is a glass tube with two metal contacts. A small quantity of mercury in the tube can bridge the two contacts to complete a circuit, or if the tube is tilted, can flow to one end and break the circuit.

To keep the mercury clean and minimise arcing the tube is filled with hydrogen and sealed. The two contacts are connected to stranded flexible conductors. The switch may be mounted on a rocking carrier which is tilted by the sensitive element; or it may be mounted directly on a bi-metallic strip as is the case in some room thermostats. This type of thermostat has a long bi-metal strip wound into a spiral. The inner end is fixed to the thermostat body and the outer end carries the mercury switch. Change of temperature causes the spiral to wind or unwind and tilt the mercury switch.

Mercury switches are very sensitive and quick acting as mercury is very mobile.

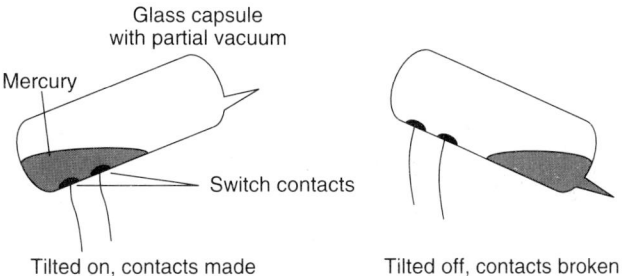

Accelerator

Air is a bad conductor of heat. As a consequence of this the temperature monitored by a room thermostat will lag, particularly if the room temperature is rising fairly rapidly. In order to overcome this disadvantage a small resistor is placed inside the thermostat case. When the thermostat calls for heat, a current flows in the resistor and generates a little heat. In this way the thermostat can keep up with or even get ahead of room temperature.

This "accelerator" resistor may be connected in series or parallel. When connected in series the whole of the current through the thermostat flows through the accelerator which has a low resistance. When connected in parallel it is across the circuit switched by the thermostat. A parallel connected resistor has a high value of resistance. Many thermostats have the mechanism sealed to prevent tampering and the repair of defective thermostats is not easily achieved and is not an economic proposition. Defective thermostats are usually replaced.

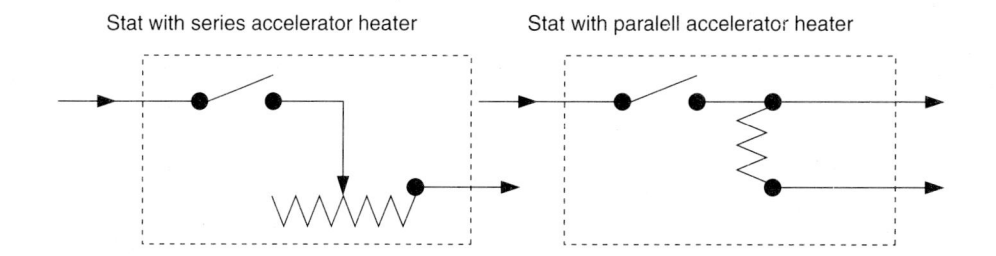

Stat with series accelerator heater   Stat with paralell accelerator heater

Application of Thermostats

Room Thermostats

The standard method used for controlling the temperature of the heated space is by controlling the output of the heater. Opening or closing the contacts is controlled by the temperature sensitive element. Room thermostats should be mounted in a position which corresponds to average room conditions. It should not be on an outside wall, near a door or window, near a source of direct heat or in a draught. The usual position is on an inside wall between 1.2m to 1.7m above floor level.

Cylinder Thermostat

This is a heat-operated electrical switch used to sense the temperature of the water in a domestic cylinder and, in conjunction with a diverter or motorised valve, may be used to control the domestic water temperature. It may be used to give priority for heating domestic hot water needs in preference to space heating demands.

Although temperature adjustment may be provided, a cylinder thermostat is normally set to cut off at 60°C (140°F). A large built-in differential, some 11°C (20°F) ensures that rapid on-off operation of the boiler does not occur at times of light loading i.e. summer operation with space heating off.

The stat is normally strapped onto the cylinder at about one third of the height from the bottom of the cylinder.

Limit Thermostats

Limit-stats operate in the same way as a room thermostat but instead of controlling the space temperature it limits the maximum temperature of hot air inside the heater. An excessive rise of temperature within the unit is detected and the unit is switched off. The unit will remain off until such time as the internal temperature falls, when the limit thermostat will allow the unit to become operative. This limit control may have an "automatic reset" or may require manual resetting after an overheat shut down.

Frost Thermostat

A frost thermostat can be wired to bring on the heating system, overriding any other controls should the temperature fall below a set point, usually -2°C or -3°C.

It is often difficult to select the best position for a frost thermostat but it should be placed at the coldest point of the system.

Fan Thermostats

These are similar to both the room and limit thermostats in operation but have a different function. The fan thermostat senses the hot air inside the heater and switches on the main fan. This control is necessary because when the heater is first started up, the heat exchanger is quite cold. If the main fan is allowed to run at this stage cold air is blown into the space calling for heat. Having the fan thermostat fitted into the fan motor circuit, the main fan is not able to start until the heater has raised the air in the exchanger to the required operating temperature. Some units are fitted with an automatic/manual switch which can be used to override the fan thermostat and enable the fan to be used for circulating cool air during hot weather.

Clock Controllers

A clock controller is a device to switch an electric circuit on and off automatically in a time sequence chosen to meet the needs of the user. Although there are several makes of clock the basic principles are similar for each. The control consists of a dial face, control knobs, a number of 'on' and 'off' switch levers and a time indicator. The dial is divided into 24 hours or may be numbered from 1-12, twice, either coloured differently or marked "Day" or "Night" or both.

The basic clock controller consists of a 240V synchronous motor, self-starting and internally fused. It can be fitted with ON and OFF switch levers to vary the number of 'on' and 'off' periods per 24 hours. Several types are made, most have three or four pin connections.

Current rating is normally 5 to 30 amps. They are usually provided with a switch to override the clock contacts and give continuous operation if required.

With a four pin time switch the clock switch connections are electrically separate from the clock motor mains supply. Extra low voltage circuits are switched by the use of this type of time clock.

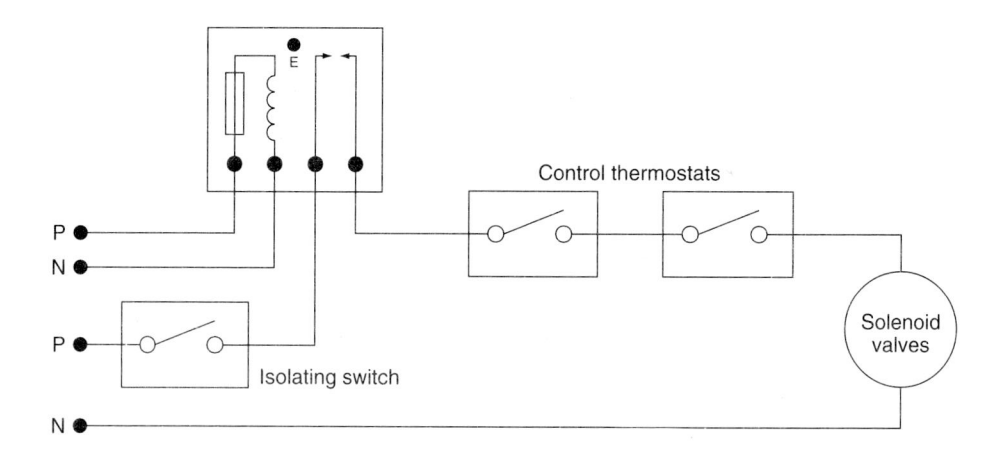

Programmers

Programmers include the basic clock unit and switch with multiple contacts which give the user the choice of an appropriate programme of heating and/or domestic hot water supply. The selection may be by push button or control knobs with various types of priority switches. The programmer has a terminal block for connection to the air thermostat, boiler thermostat, gas solenoid valve and circulating pump.

A programmer usually has a number of manually operated switches, for example:

One switch for central heating
One switch for special purposes (e.g. zone heating or water heating)

Each switch usually has three positions - OFF, CONSTANT and TIMED.

More advanced programmers use a multi-positioned switch which gives a wider choice of hot water demand and heating demand. A typical six choice programmer can give the following:

1. OFF
2. Hot water on twice - central heating off
3. Hot water on twice - central heating on twice
4. Hot water on all day - central heating on twice
5. Hot water on all day - central heating on all day
6. Hot water and central heating constant

If the clock controller or programmer is not integral with the boiler, it should be placed in an easily accessible position.

The repair of these devices is usually beyond the capabilities of a service engineer apart from changing a blown fuse.

Electronic Programmers

There are two types generally available which fall into the following categories:

LED (light emitting diode) display
LCD (liquid crystal) display

The method of setting is similar to setting a clock radio and if they do go faulty the entire unit should be replaced.

With some types, the clock is battery powered and a new battery should be tried before assuming there is a fault.

Electrical Components

Relays

A relay is an electromagnetic device arranged so that a current in one circuit can switch the current of another circuit e.g. switching a mains voltage circuit in response to a low voltage circuit or vice versa.

The relay consists of an electromagnet and moving armature which can operate a switch. When there is no current in the coil, the armature is held away from the magnet by a spring. This gives one switch position. When current flows in the coil, the armature is attracted enough to overcome the spring and operate the switch.

The switch used is generally a simple fixed and moving contact. The moving contact may be carried either on the armature or mounted on a flat spring, which is deflected by the armature. The relay can operate a number of switches which are insulated from each other.

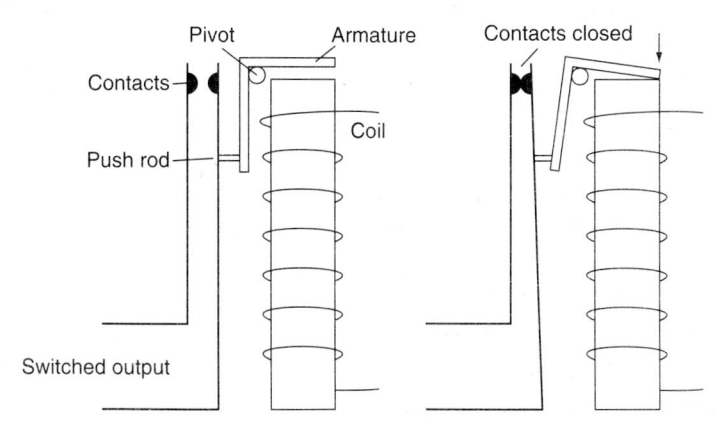

Each switch can be arranged to be normally closed or normally open. The type most likely to be found will have an extra low voltage coil and one or two normally open contact switches. This type of relay is used for switching a mains voltage pump via an extra low voltage thermostat circuit.

Common Faults and Causes of Failure

Mechanical failure - due to dirt in the mechanism or a weak or broken spring.
Electrical failure - open circuit or short circuit due to vibration, excessive temperatures or insulation failure.

Solenoid Valves

Electro-magnetic devices can be used to operate gas valves. A simple coil producing an electro-magnetic force is used to attract a soft iron 'armature'. This in turn acts directly on the gas valve.

In the case of the device illustrated the valve is normally held closed by the spring. When the coil is energised, the armature is attracted, overcoming the spring and opening the valve. It is assisted by a small fixed pole-piece situated at the top of the tube and will remain open until the coil is de-energised when the valve will snap closed under the pressure of the spring.

Solenoid valves are very robust and reliable in use. If they are suspected of being faulty they should be replaced.

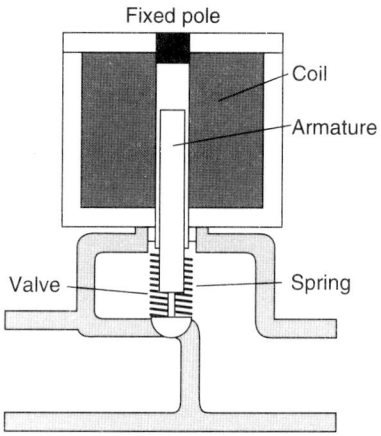

Central Heating Pumps

The function of a central heating pump is to circulate hot water through the system. In doing so it may have to overcome the force of gravity in raising the hot water to higher level, and also the resistance to flow offered by the installation.

The design of the glandless pump is such that the entire rotating element is completely separated from the non-rotating parts by a shell of non-magnetic metal. The bearings are of the water lubricating type requiring no grease or oil, so absolute water tightness is assured. Materials of construction in contact with the water are corrosion resistant.

The motor and impeller are both mounted on the same shaft, there are no complicated and expensive seals, fewer bearings and moving parts and less chance of leakage. The pumps are simple, cheap, reliable and very compact.

The impeller comprises a number of vanes which rotate within a closed circular chamber. Water enters the 'eye' of the impeller and is 'impelled' by centrifugal force to the periphery, where the outlet is situated.

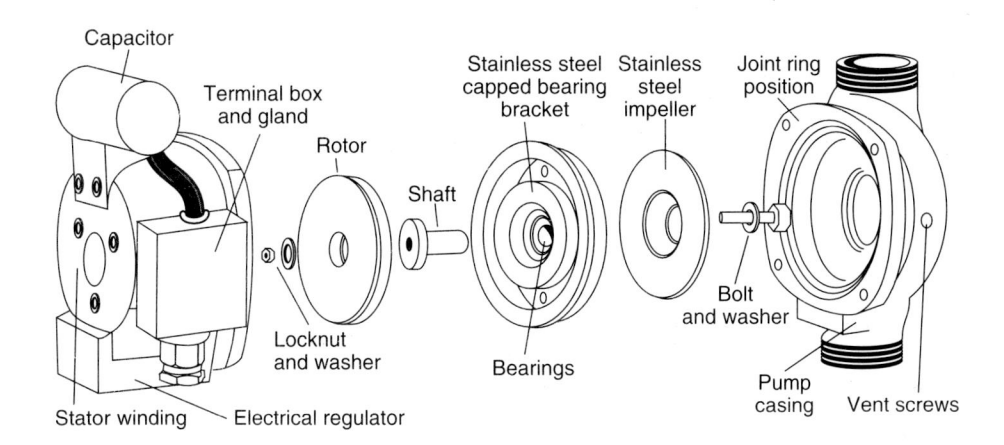

Pumps should always be fitted where they are accessible for exchange, inspection and maintenance. Whenever a system is fitted with controls which can close off all flow (such as thermostatic radiator valves and a cylinder thermostat) there must be a bypass fitted between the flow and return pipes to ensure that water can flow through the boiler and pump at all times.

Fans

Electrically driven fans are to be found in association with almost every conceivable type of gas installation.

Most of the fans encountered by the service engineer are small low powered units used to perform one or more of the following functions:

- to draw fresh air into an appliance, or the environment of an appliance
- to provide air for combustion (fan-assisted balanced flue)
- to circulate the air or heated air within an appliance
- to distribute the warmed air from an appliance to the environment (warm air central heating systems)
- to assist in the removal of combustion products to atmosphere (fan-assisted open flue)

There are two main basic types of fan used in conjunction with gas appliances, the propeller type of axial flow fan and the centrifugal fan.

Propeller Fan

The propeller or airscrew fan has multiple blades which radiate out from the axis of rotation.

These blades are shaped so that rotation of the fan propels the air by a screw action along a path parallel to the axis of the fan (axial flow). A properly designed fan blade has a complex shape varying along its length in width and 'angle of attack'. A fixed shroud is usually placed around the periphery of the fan to direct the flow and to prevent air from flowing back round the outside of the fan. If permitted this back flow of air will reduce the fan's efficiency. Most fans used in conjunction with gas appliances and warm air systems have these shrouds or volutes and closeness of fit to the outside edge of the fan is important.

Small propeller fans usually have the fan blades mounted on the motor shaft. This design means that they are compact and the airflow must pass over the motor. This aids cooling but if used to move dirty or greasy air in the fan motor can quickly become soiled.

There are several methods of adjusting the volume of air passed through the fan. The most common system is to vary the speed of the fan motor electrically. This may be done by a simple two speed control, or by variable electronic speed control. Mechanical systems include restricting the effective aperture by means of an iris or a louvre shutter mechanism.

Propeller fans are simple, reliable and robust, but they can be noisy, particularly when running at a high speed.

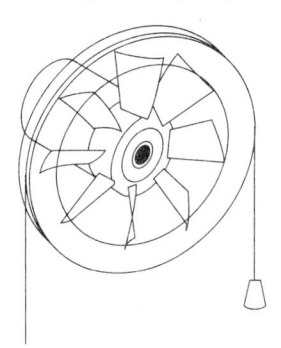

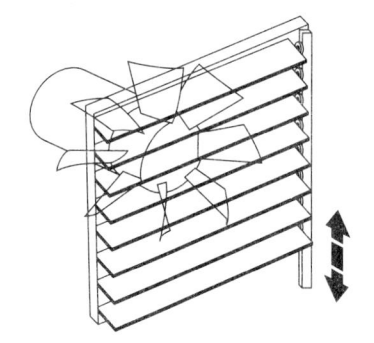

The Centrifugal Fan

The alternative fan design commonly used is the centrifugal or 'paddle wheel' type. In this design the fan blades are positioned equidistant and parallel to the axis of the fan rotation. The blades may be straight, forward or backward curving or of 'aerofoil' section. The blades extend inwards from the circumference leaving an unobstructed central hollow. As the fan starts to turn, air is pushed outwards by the blades and a zone of reduced pressure is produced in the central hollow. Air is drawn into this zone at the ends of the fan. If the fan is direct drive the opposite end of the fan from the motor drive end is left open to allow air to flow into the fan.

The air flow through the fan is radiated out equally from all points around the periphery of the fan and to control the air flow the fan has to be mounted in a housing that will channel the output where desired.

In comparison with the propeller type there is an even pressure distribution around the fan so a greater pressure difference across it can be generated. This allows the fan to operate efficiently where restrictions to the air flow exist.

Centrifugal fans can be designed for a variety of applications. A wide fan with short blades will shift a large volume of air but will only give a small pressure rise. Such fans are suitable for fanned convectors with their short length of enclosed ducting (over the heat exchanger). A narrower fan with longer blades is necessary where higher air pressures are required such as in a ducted warm air system. The construction of centrifugal fans, with the motor outside the air current, makes them suitable for moving hot gases.

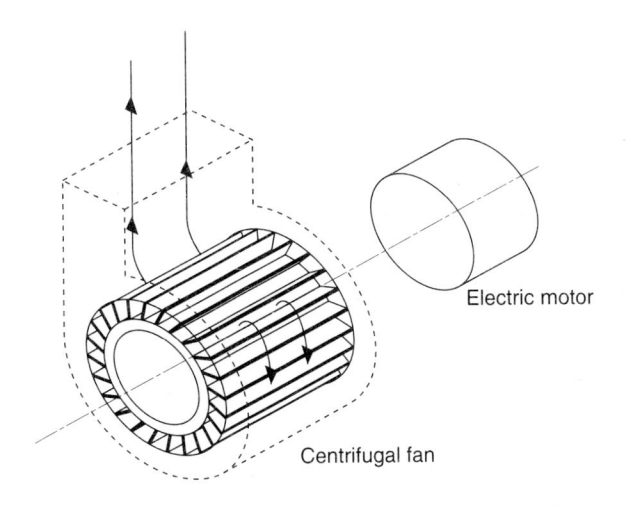

Electric motor

Centrifugal fan

Motorised Valves

A motorised valve is one which is actuated by an electric motor. There are a number of ways in which this can be done.

In its simplest form a small motor drives a valve controlling the water flow. When the electricity supply is switched on, the motor runs, turning the valve through a quarter of a revolution, or 90° to the 'on' position. At this point the motor changes-over a switch contact and breaks the circuit. The valve remains in its new position until the electricity supply is again switched on to the motor by the control circuit, when the motor turns the valve another 90°, shutting it off.

The valve continues to rotate 90° each time the motor runs, stopping alternatively at the 'on' and 'off' positions.

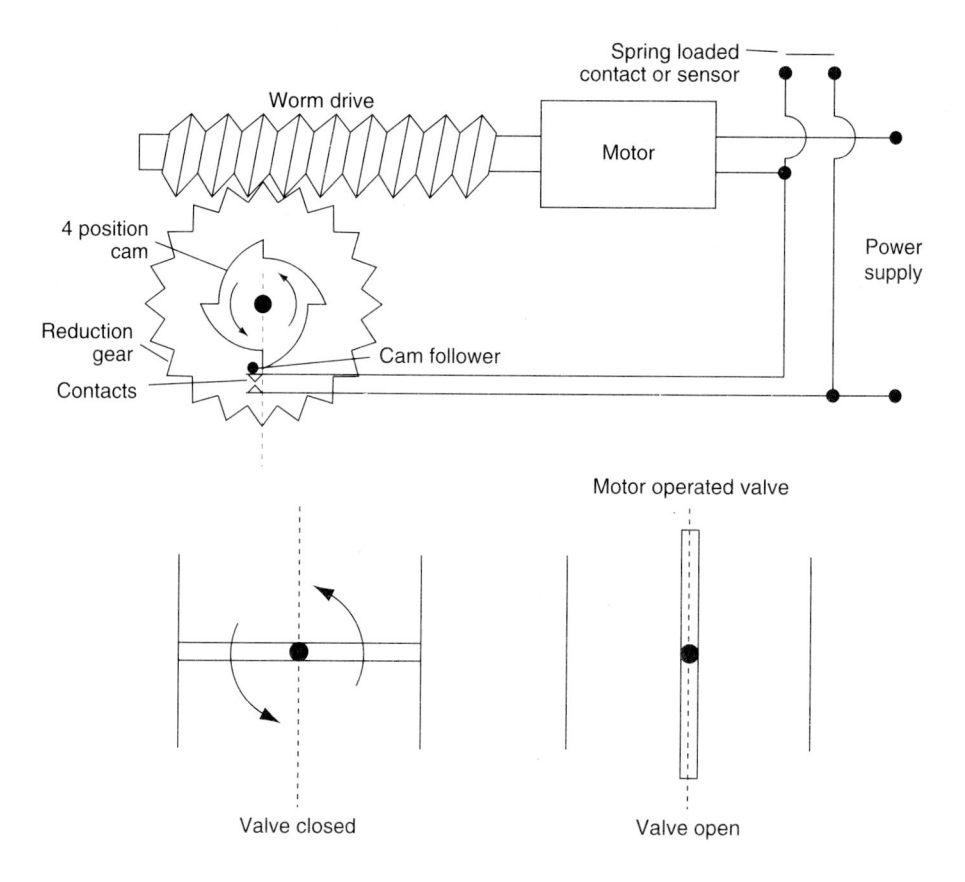

Diagram: Motorised valve mechanism showing worm drive, motor, spring loaded contact or sensor, 4 position cam, reduction gear, contacts, cam follower, and power supply. Below: motor operated valve in closed and open positions.

In some motorised valves the action of opening the valve compresses a spring which closes the valve when current is switched off.

Valves are used for a variety of purposes. Some which give on-off control have a simple plug, or a rotary vane, called a 'butterfly valve'. Other, more complex designs can be used as two way valves to change over the circulation of water in a central heating or water heating system. These are known as zone valves.

Ignition Systems

A pilot light is a small flame, located so that immediately the main gas supply is turned on, it will be ignited by the pilot flame. The small gas supply necessary for the pilot is taken off the main supply, ahead of the control valve.

The pilot is often combined with a thermal cut-off flame failure device.

The ignition of the pilot flame can be achieved using either of the following methods:

- filament ignition
- spark ignition

Filament Ignitors

The ignitor consists of an electrically heated filament which can either be mains or battery operated, placed adjacent to the pilot flame gas jet. The filament is a small coil of thin resistance wire, usually of platinum.

The power required for heating the filament is around 3V. This can be obtained from batteries or from a step-down transformer. The transformer has the advantage of consistent power output and avoids the need to change batteries. An ignitor head incorporating a lighting jet is illustrated.

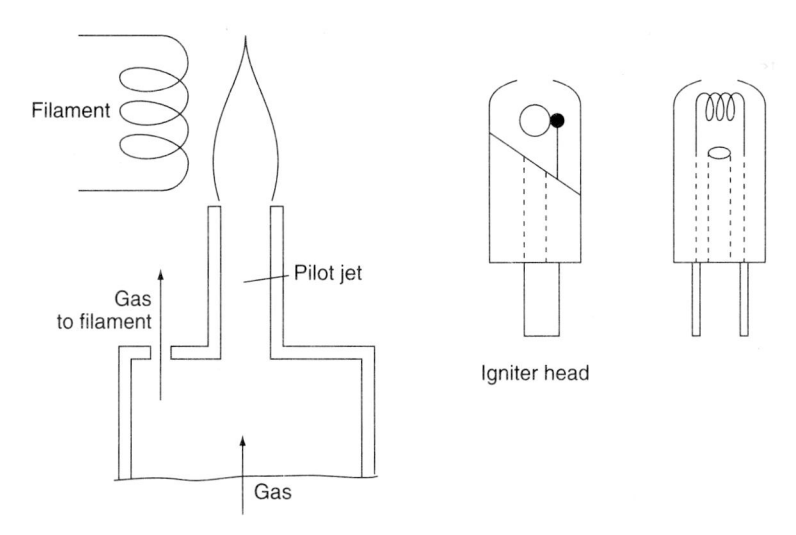

The protective shield which surrounds the filament must permit air to the jet to obtain a combustible mixture.

Glow coil ignitors have a heavier filament and are heated by mains electricity through a step down transformer which provides a voltage of 2.5V or greater.

Spark Ignition

Gas may be ignited by a high voltage spark of between 5,000V - 15,000V. These voltages are obtained from:

piezo-electric generator
mains transformer
electronic pulse system

Piezo-Electric Ignitors

Lead zirconate - titanate crystals which are used in these ignitors are exposed to powerful electric fields during manufacture which polarise the material. When the crystals are stressed or deformed they produce an emf of about 6,000V between the two ends of the crystals.

The ignitor usually consists of two crystals each about 12mm long and 6mm diameter connected in parallel. The crystals are separated by a metal pressure pad which is connected to the spark electrode supply and the other end of the crystals are earthed to the casing as illustrated.

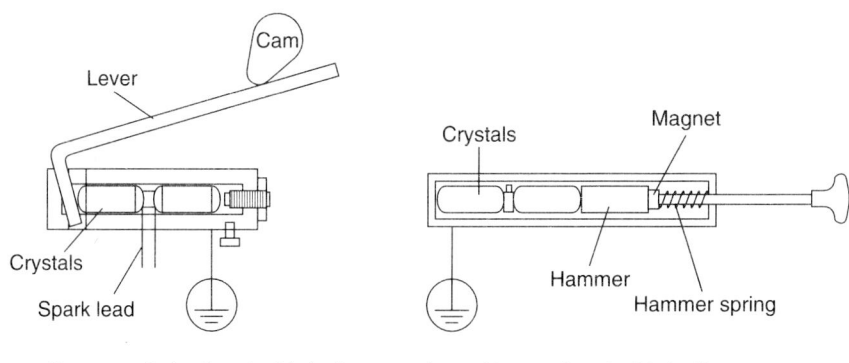

Cam operated peizo electric igniter   Impact type peizo electric igniter

Depressing the lever B applies pressure to the crystals, so the voltage builds up until it overcomes the resistance of the spark gap and a spark is produced; another spark is produced when the lever is released and the stress removed. The lever can be operated in a number of ways. In the case illustrated, it is by a cam, which would be attached to the top spindle. It is also possible to apply pressure to the crystals by impact.

Mains Transformer

The ignition transformer is a step up transformer with the electrode connected between the ends of the secondary winding as illustrated to give a voltage of between 5,000V to 10,000V.

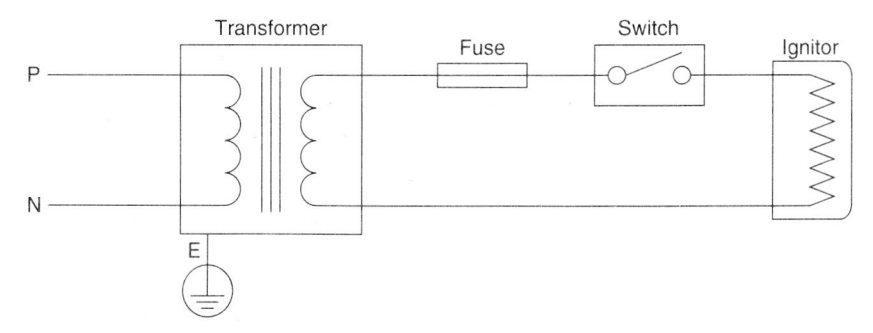

On higher voltage systems, the secondary winding is earthed from a centre tap to keep the potential between either output connection and earth down to half the output voltage.

Electronic Pulse Ignition

With the development of solid-state devices it has become possible to produce sparks at speeds of up to eight sparks per second at several electrodes simultaneously from a single, small generator. Although the voltage of the spark may be 15,000V the actual danger from contact with the electrodes is negligible.

These devices can produce their sparks by feeding a current through a rectifier into a capacitor as illustrated. The capacitor is then discharged through a small step-up transformer, producing a very high voltage in the secondary winding, which is connected to the electrodes.

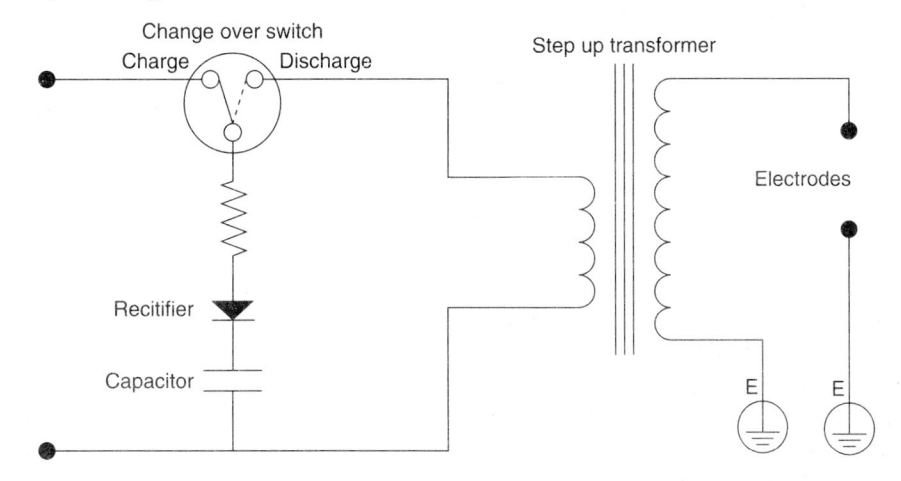

The changeover switch has a 'charge' position on contact 1 and a 'discharge' position on contact 2. It could be operated manually by a push-button. On more complex ignitors the discharge may be brought about electrically. Solid-state oscillators may be used in conjunction with silicon-controlled rectifiers to provide the pulse. The power supply is usually mains voltage.

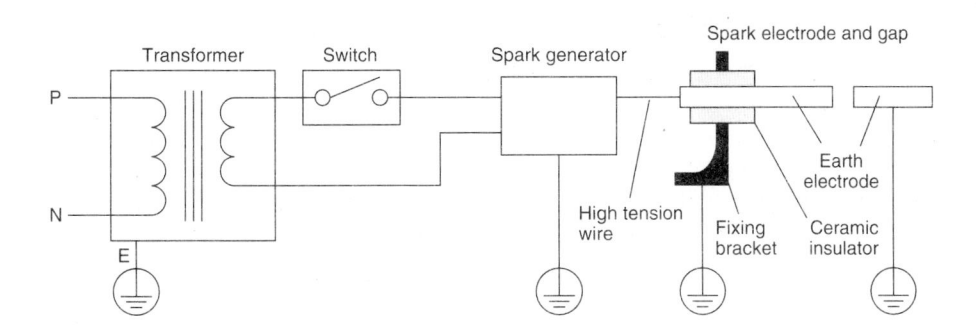

Common Causes of Failure

Filament Ignition Systems

Failure of the ignitor to light up can be due to lack of air at the filament, blockage of the jet or gas supply, broken filament or failure of the electrical supply.

Spark Ignition Systems

In all the spark ignitors the position of the electrode is critical. Not only must the gap be correct but also the spark must occur in a place where the air/gas mixture is well within the flammability limits. The speed of the mixture must allow the flame produced to light back onto the flame port.

The insulation must be very good to prevent leakage at the high voltages produced. On piezo-electric crystals, dampness on the crystal assembly may result in the current shorting-out to earth with consequent loss of spark at the electrodes. When faults occur within the spark generators themselves, the entire unit should be renewed.

Circuit Diagrams, Wiring Diagrams and Cabling

A circuit or wiring diagram is simply a means of conveying information about an installation or appliance more easily or clearly than can be expressed in words.

Perhaps the simplest form is the Pictorial Diagram which provides a easily recognisable illustration of the various componenets and their connections. This type of illustration does not show the electrical relationship between components or the layout of the system, how the circuit works or the most convenient way of wiring the circuit.

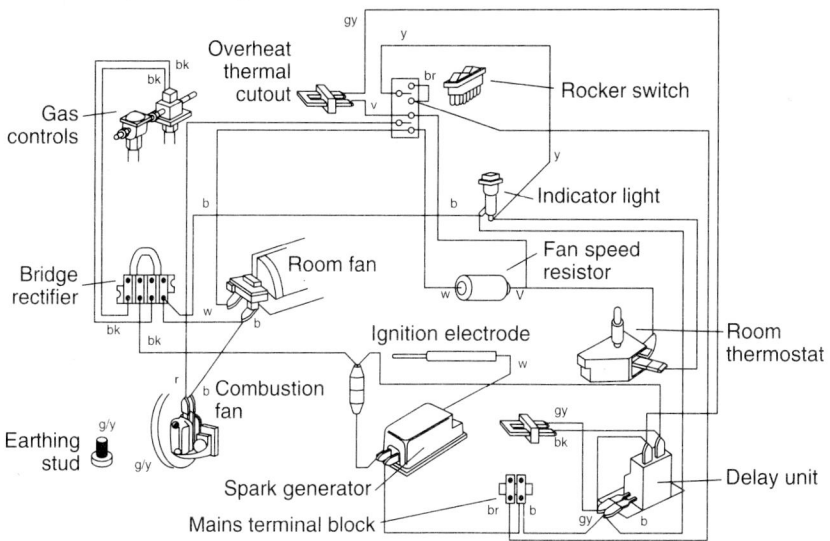

A circuit diagram uses standard symbols for the various elements and is intended to illustrate how the installation functions.

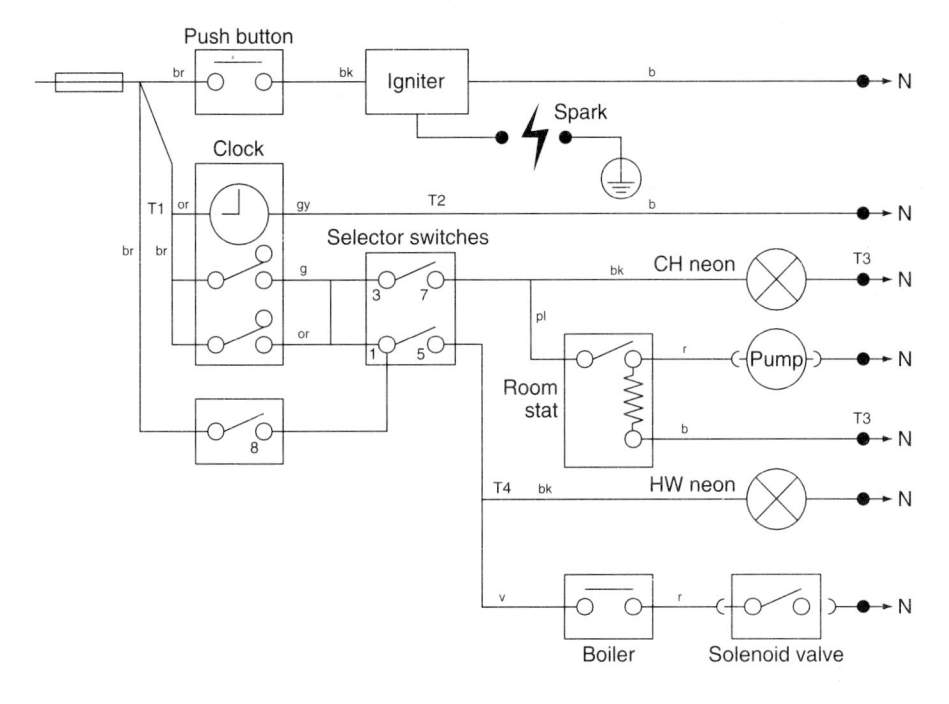

Wiring diagrams provide information about how the various components are interconnected. They may specify wiring by a colour code, or with numbers or symbols. A wiring diagram may not necessarily convey information about how the circuit operates, or the optimum layout for the various components.

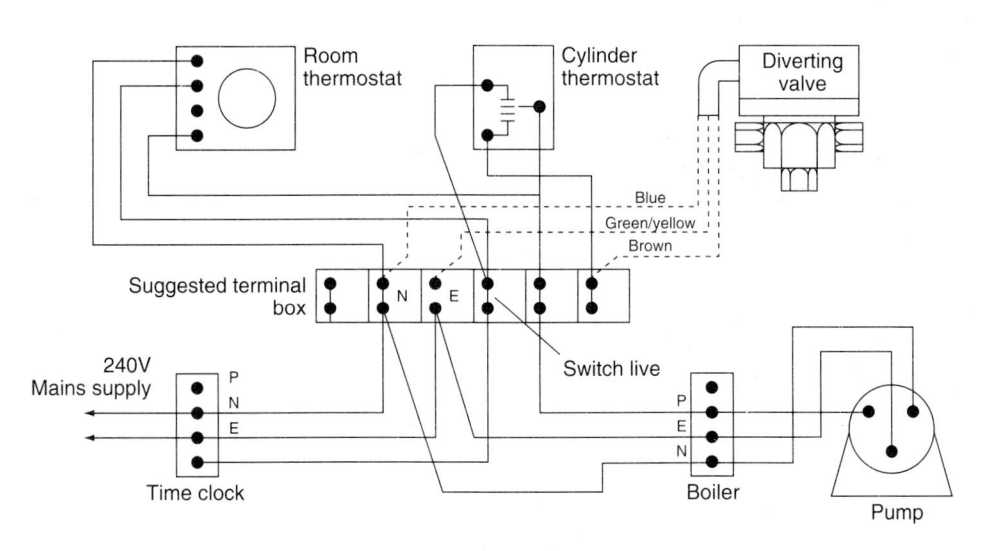

Cabling diagrams shows the cables required for the interconnection between components of an installation and will specify the type of cable, number of cores, earth connections, etc.

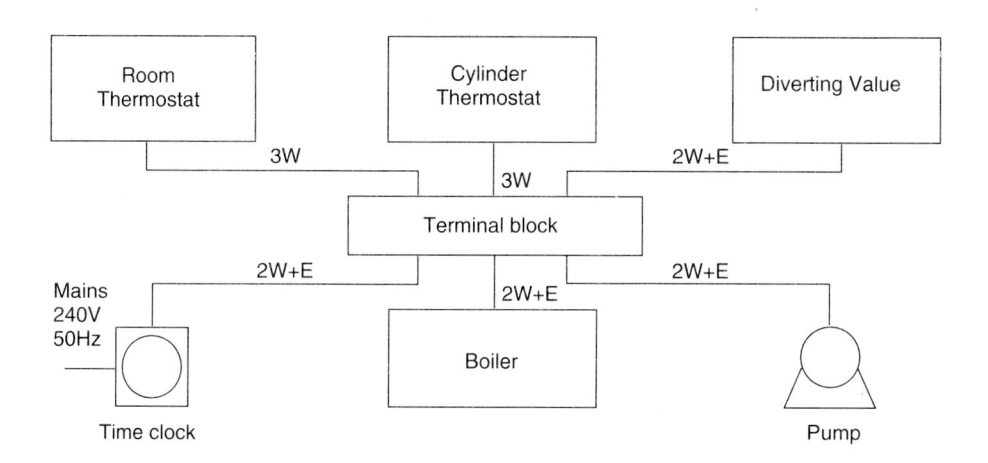

Layout diagrams show the physical layout of the installation in pictorial or symbolic form.

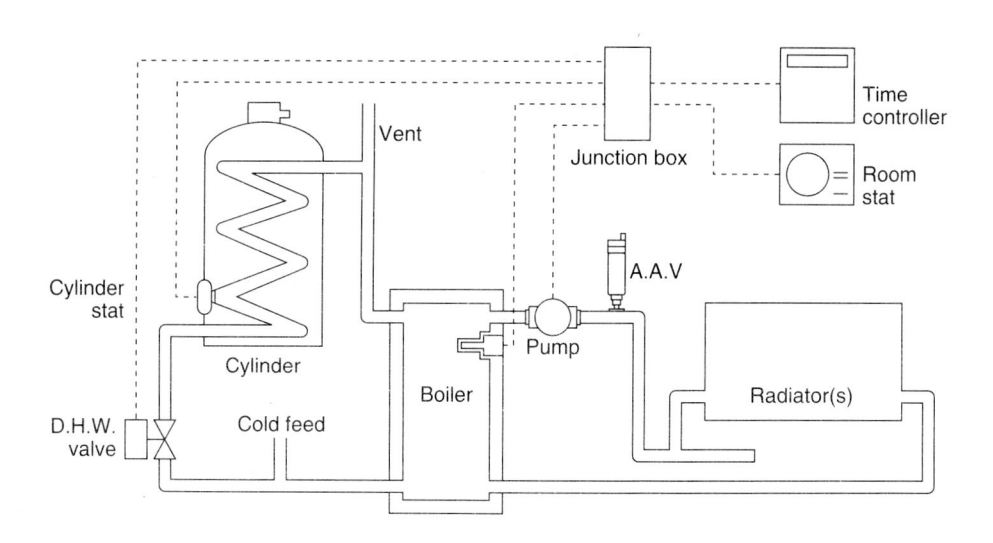

Electrical Symbols

All types of diagram may make use of standard symbols and it is important that the gas fitter should be familiar with these. Some of the most common are listed below:

Battery	—┤├—	Glowcoil	•⋀⋀⋀⋀•
Resistor	—⋀⋀—	Thermostat	—[t⁰]—
Resistor (variable)	—⋀⋀—	Indicator lamp	—⊗—
Capacitor	—┤├—	Electric motor	—(M)—
Capacitor (variable)	—⫽—	Pump	—(P)—
Inductor	⌒⌒⌒	Fan	—(F)—
Transformer	⌒⌒⌒/⌒⌒⌒	Electricity meter	▢
Rectifier	—▶⊢—	Distribution board	▢
Solenoid (relay)	⌒⌒⌒ • •	Main switch	▢
Terminal	—•⊢—	Main control unit	▢
Earth connection	⏚	Switch	—o⁄o—
Spark electrode	—•↯•⏚	Socket outlet	—(•—
Flame electrode	—• •⏚	Plug and socket connector	—(▬—
Fuse	—▭—		

Control Systems

Elements of Control

Imagine a man sitting alone in a room with an old fashioned paraffin heater. Towards evening, as the temperature drops, he feels cold and lights the heater with a match, turning up the wick to do so. After a while he feels warmer and adjusts the heat output by turning the wick down; or he may extinguish the flame altogether.

This simple analogy illustrates the basic elements of a control system. What is being controlled is the temperature of the air in the room. The sensor is the man himself, who reacts to his environment; he senses whether he is too hot or too cold and acts to control the condition. The controlled device is the heater and the controller, the wick which can be adjusted, up or down as necessary.

Notice that three elements are necessary for a control system, a sensor, a controller and the controlled device.

Basic Manual Control

In a basic, gas fired installation the primary circulation may operate through gravity, the heated water rising being replaced by cold water. If a number of radiators are included, these might be fed by an electric pump.

If a switch is included in the pump circuit, this will provide a mechanism for controlling the period of time that the radiators are heated. A switch would also be required to isolate the installation from the electricity supply and the circuit would need to be protected by a fuse. Physical controls (stop cocks) might be used to control the flow of heated water to individual radiators but the boiler and pump would operate continuously unless switched off.

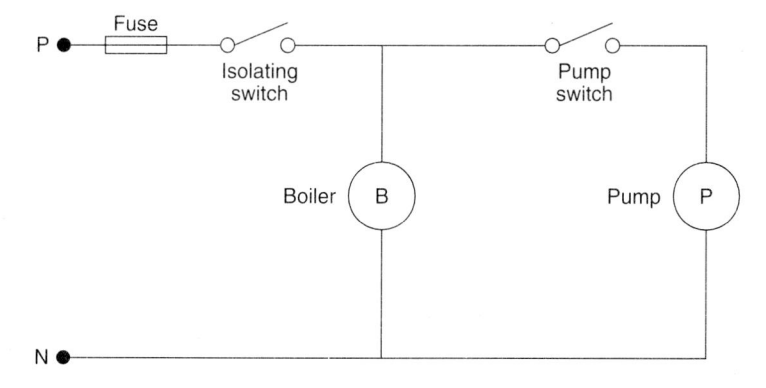

Such a system would be inconvenient and expensive to operate. There are a number of devices which can be used to control the installation more efficiently and automatically. These include mechanisms for controlling temperature and the period(s) of operation (time).

Time Control

There are now a great many timing devices available which can be employed to switch on (or off) an electric circuit. The simplest of these includes a clock mechanism which operates a switch at a pre-set time.

Devices such as this might be used with the basic installation discussed previously. For example, it might be used to switch on (or off) the boiler or the pump at predetermined times to meet the needs of the user.

More sophisticated programmers can be used to "switch" separate parts of the circuit, for example, heating in one room only, or only on certain days. The basic principle of operation is the same - the mechanism makes (or breaks) a switch which controls power to an electrically operated component.

Temperature Control

Thermostatic controls of various kinds can be used to "sense" temperature at different locations and to operate a variety of controls to switch on (or off) a heating (or cooling) sequence. In practice a wide range of temperature sensitive and timing devices are used together to provide an efficient control over the operation of the installation, for the benefit of the user.

A simple circuit, providing control over both temperature and time is depicted below:

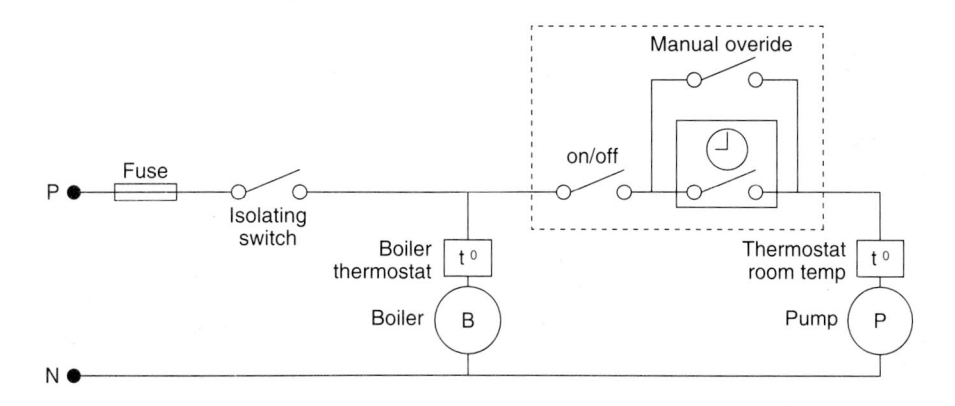

A simple arrangement such as this will afford a considerable degree of control. The "timer" can be set to switch on the circulating pump at times when heat is likely to be needed; for example, first thing in the morning for, say, two or three hours and again in the evening; switching off at bedtime. A manual override switch will normally be provided to provide a measure of flexibility, bypassing the timer; and another switch permits the user to switch the circuit off entirely, isolating the pump and associated controls from the supply. Thermostats provide additional control, sensing water temperature and room air temperature. These will be set at predetermined levels and are so arranged as to complete the circuit when heat is called for.

High/Low Limit Control

One disadvantage of this simple system of control is that during the "off" periods the air temperature may fall too low, particularly in extremely cold weather, and it would require a great deal of heat (energy = cost) to raise the temperature to the required level when the systems switches "on". To prevent this happening, a limit control is fitted. This is a thermostat set approximately 10° - 15° below the desired temperature and is arranged to switch on the system, bypassing the time controller when the temperature falls below this level.

Opinions differ as to the effect on operating costs but there is no doubt that a house or other premises maintained between fixed temperature limits will be much more comfortable and factors such as condensation or frost damage will be avoided.

Short Cycling

Another factor is the effect of "short cycling". The boiler flue temperature will be higher than that of the incoming air, which will cool the heat exchanger to a point when the boiler thermostat will call for more heat. Also the water circulated by gravity will result in cold water being drawn through the boiler, which again will call for more heat. As a consequence the boiler will "cycle" to replenish its own heat losses. This is both wasteful and uneconomic.

The fitting of a cylinder thermostat will enable the boiler to operate in response to genuine demands for heat, either from the air temperature or from the temperature of the water in the cylinder. A cylinder thermostat can be inserted in the circuit between the boiler and the thermostat measuring air temperature. This arrangement allows the boiler to fire only when either the cylinder thermostat or room thermostat calls for heat.

Independent Temperature Control

So far, the temperature of the hot water system has been governed by the boiler thermostat only, but greater economy and efficiency can be provided by operating the heating and hot water systems separately.

In order to provide independent temperature control of heating and hot water supplies a motorised valve can be inserted in the primary gravity circuit. When the cylinder thermostat is satisfied this valve will be closed and even though there is a demand by the space heating thermostat, the temperature of the hot water supply will not exceed the cylinder thermostat setting.

The electrical circuit of the basic system so far described is as follows:

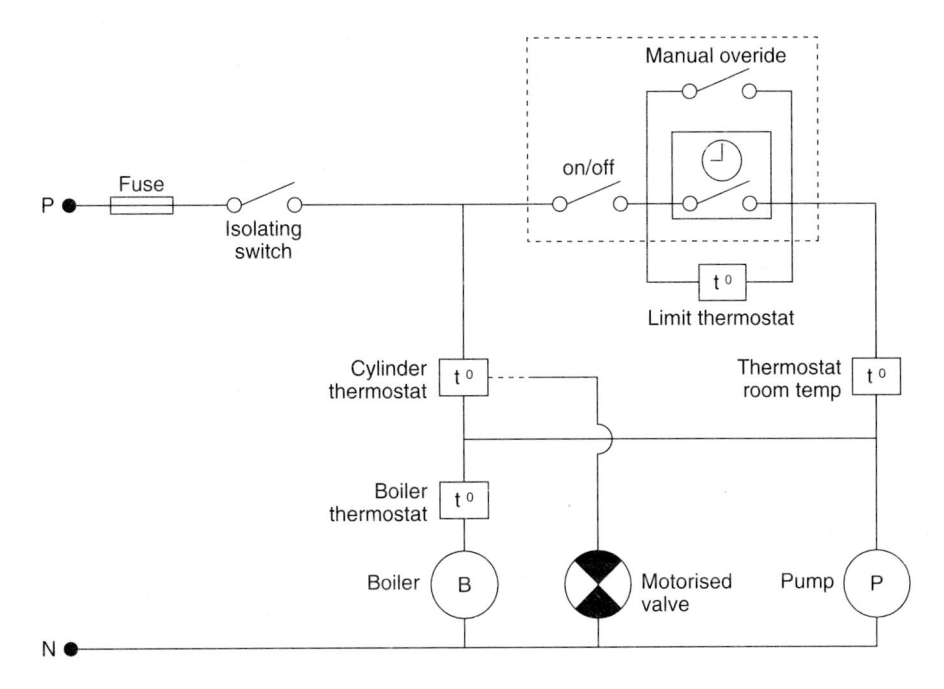

The physical layout of a basic circuit such as this might be as illustrated in the pictorial diagram shown below. Note that the three thermostats, (the room thermostat, cylinder thermostat and boiler thermostat) are linked to a central control panel which will also include timer controls, manual override and limit stat.

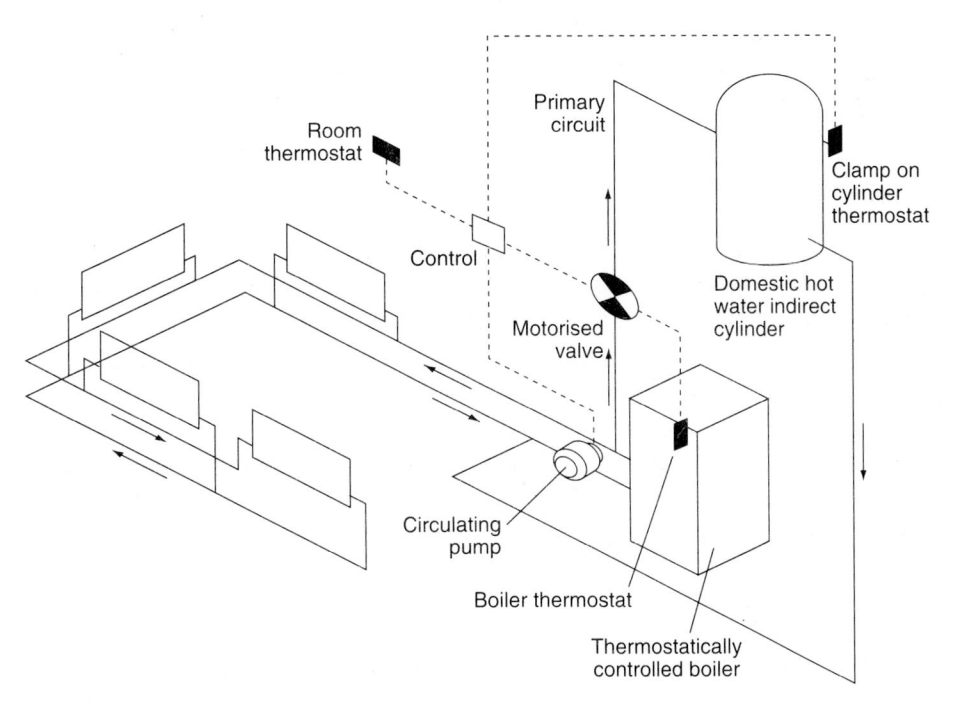

10/25

Priority Switching

In some systems it is necessary for the hot water supply to take priority over the space heating. By using a cylinder thermostat with a changeover switch the heating pump cannot operate until the hot water cylinder thermostat is satisfied. With both thermostats satisfied, the boiler will be unable to operate.

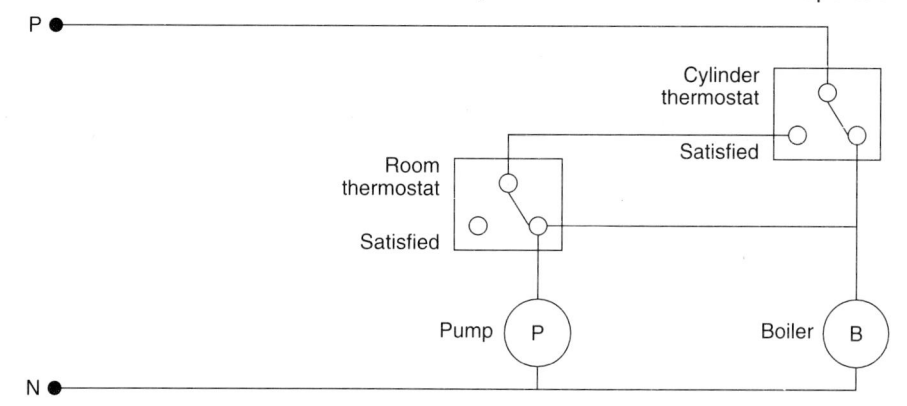

Forced Circulation of Both Heating and Hot Water Services

Another method of preventing boiler short cycling is to have the hot water circuit circulated with the pump. This involves installing a diverting valve, (which is a three-way valve) in the pipework feeding both the heating and the hot water circuit as illustrated.

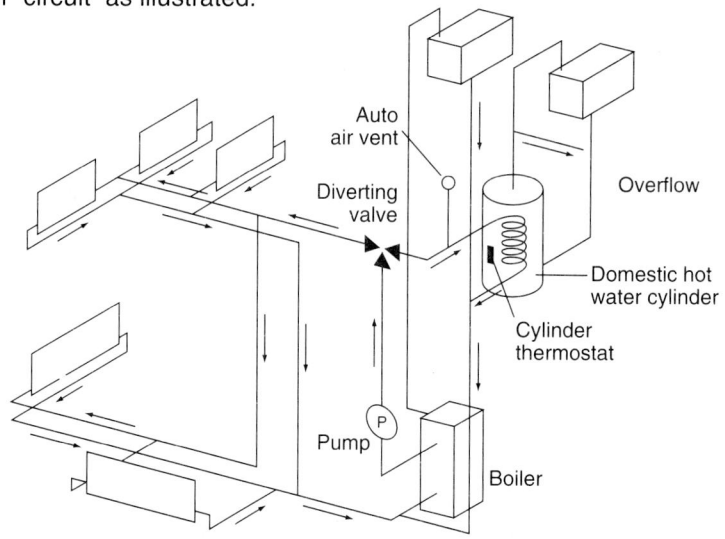

The diverting valve is governed by the cylinder thermostat in conjunction with a programmer so that priority is given to one service. When this has been satisfied the appropriate thermostat operates causing the diverting valve to shut off the water supply to that service and allow the hot water supply to reach the other service until the thermostat on that service is satisfied and shuts down the pump.

Typical Control Systems

The basic control system shown above is suitable for small domestic installations. More elaborate systems are now available using electronic programmers and sophisticated control mechanisms. Examples in common use are shown in the following pages.

The 'C' Plan

The 'C' plan is designed to provide independent temperature control of both heating and hot water circuits in pumped heating, gravity domestic hot water heating installations. Time control may be provided by a time switch or programmer.

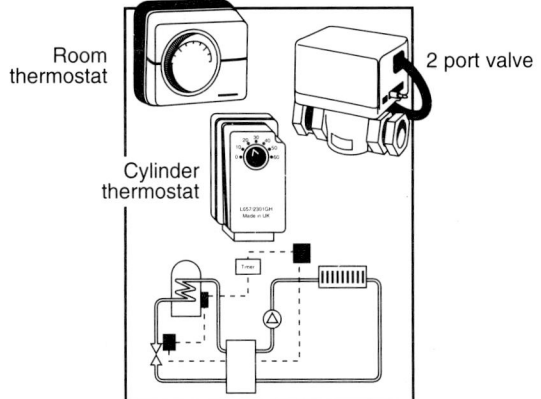

Operation

Heating Only - On a demand for heat from the room thermostat the pump and boiler are switched on. The zone value in the DHW primary remains closed.

Hot Water Only - On a demand for heat from the cylinder thermostat the valve is energised open. Just before the valve reaches the fully open position, the auxiliary switch closes and switches on the boiler only.

Heating and Hot Water - When both thermostats demand heat, the pump and boiler are switched on and the DHW valve is open.

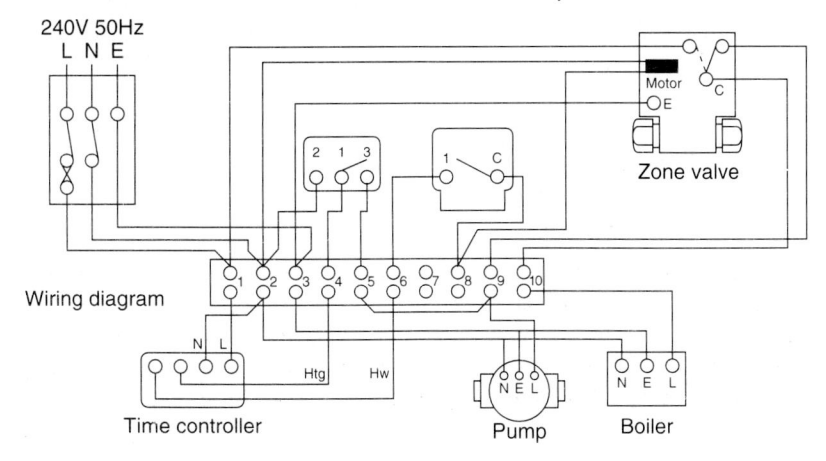

The 'S' Plan

The 'S' plan is designed to provide independent temperature control of heating and domestic hot water circuits in fully pumped central heating installations. Time control may be provided by a simple time switch or a programmer.

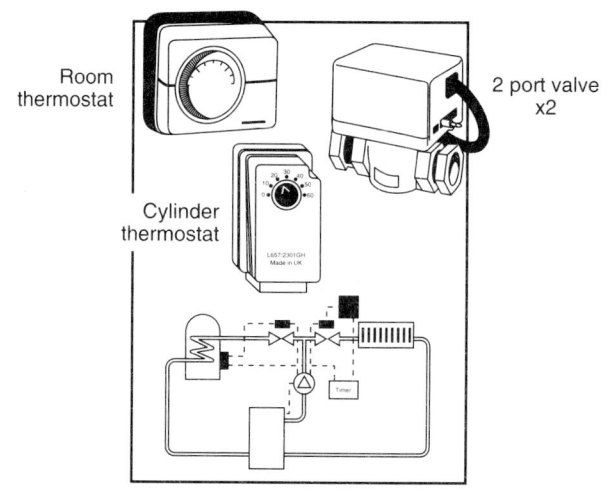

Operation

On a demand for heat from either thermostat the respective zone valve will be energised open. Just before the valve reaches its fully open position, the auxiliary switch will be closed and switch on both pump and boiler. When both thermostats are satisfied, the valves are closed and the pump and boiler switched off.

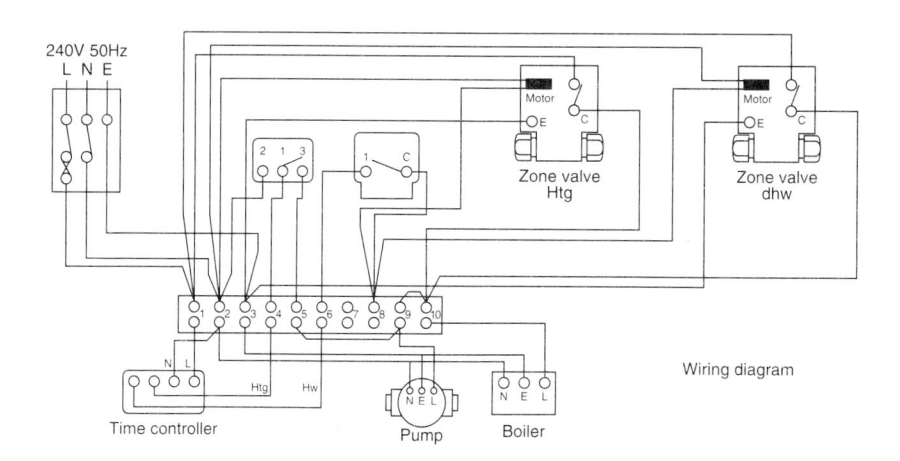

Wiring diagram

The 'W' Plan

The 'W' plan is designed to provide independent temperature control of both heating and domestic hot water circuits in fully pumped central heating installation.

The 'W' plan uses a two position diverter valve which is normally installed to give priority to the domestic hot water circuit. Because the 'W' plan is a priority control system, it should not be used when there is likely to be a high hot water demand during the heating season which could lead to the space temperature dropping below comfort level. The situation is likely to occur, for example in large family dwellings or in poorly insulated properties.

Time control may be provided by a time switch or mini-programmer.

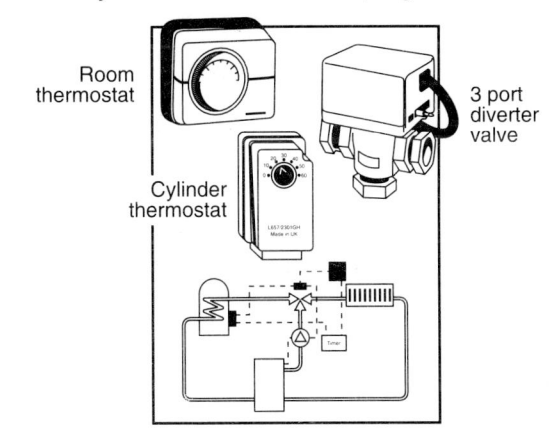

Operation

Hot Water Only Requirement - On a demand for heat from the cylinder thermostat, the pump and the boiler are switched on.

Heating Only Requirement - On a demand for heat from the room thermostat, the valve is energised, so that the hot water port is closed, and the heating port is open and the pump and boiler switched on.

Hot Water - When wired and installed to give Hot Water priority the heating will remain off until the hot water is up to the required temperature.

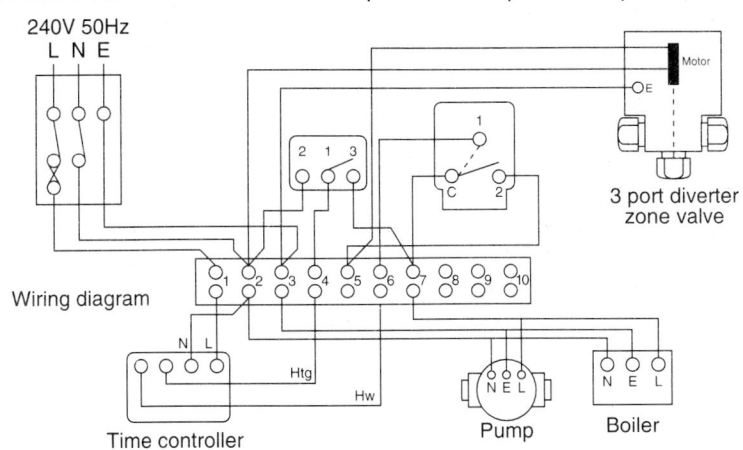

The ' Y' Plan

The 'Y' plan is designed to provide independent temperature control of both heating and domestic hot water circuits in fully pumped central heating installations. Time control may be provided by a simple time switch or a programmer.

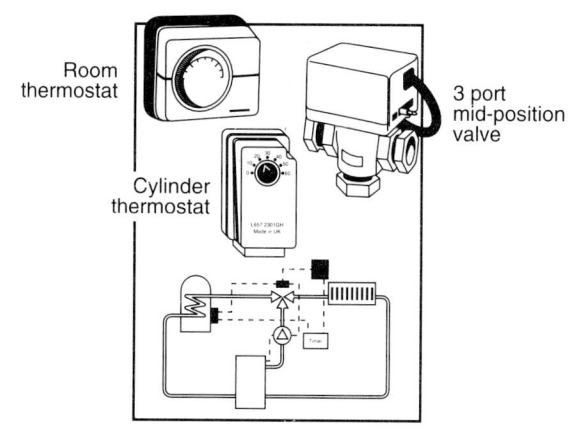

Operation

Heating Only Requirement - On a demand for heat from the room thermostat, the valve motor is energised so that the CH port only is opened and the pump and boiler switched on.

Hot Water Requirement - On a demand for heat from the cylinder thermostat, the valve remains open to DHW only and the pump and boiler switched on.

Heating and Hot Water - When both thermostats demand heat, the valve plug is positioned to allow both ports to be open and the pump and boiler switched on. When neither the thermostat is demanding heat, the pump and boiler are off. The valve remains in the last position of operation whilst the time control is in the "ON" position.

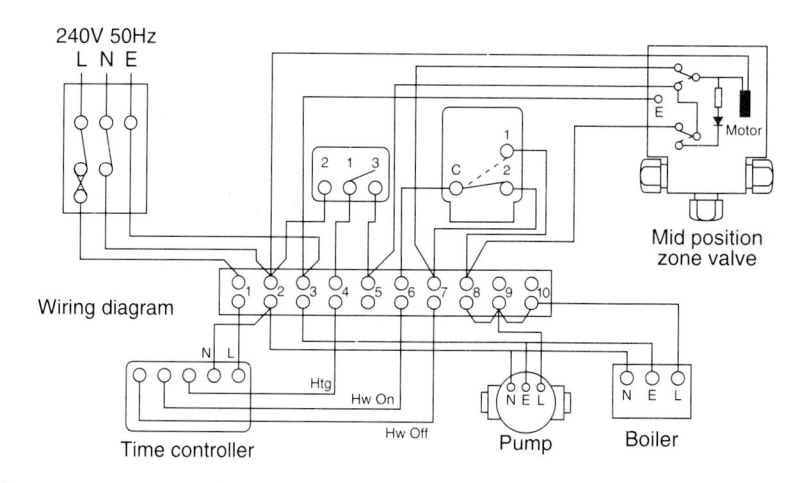

Fault Finding Techniques

Six Basic Steps to Fault Finding

The essence of fault diagnosis is logic. If a logical approach is adopted when investigating, testing and drawing conclusions, and the operative has received adequate training in related skills, the path from fault to remedy should be brief. Training should aim to develop an approach which embodies the six basic steps of fault finding.

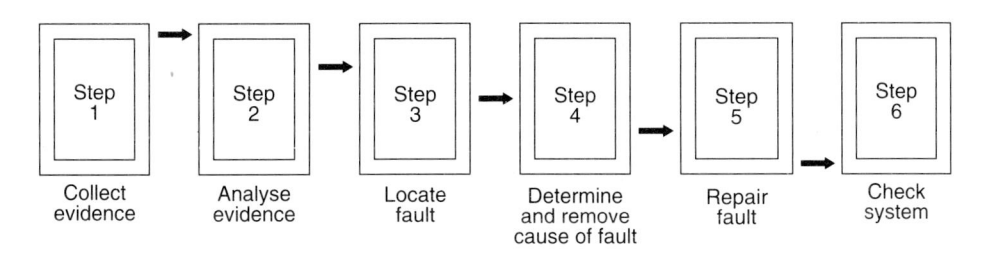

```
Step 1          Step 2           Step 3         Step 4              Step 5         Step 6
Collect         Analyse          Locate         Determine           Repair         Check
evidence        evidence         fault          and remove          fault          system
                                                cause of fault
```

If this process is followed and the six steps linked together in a logical sequence, they provide the basis for any fault diagnosis.

A fault or defect is normally considered as the failure of any component or system to perform at its predetermined level of efficiency - thus causing a 'problem'. On domestic heating and gas appliances the definition of a fault includes anything which prompts a customer to complain.

Flow Charts

The following flow charts indicate a logical approach to fault finding.

Electricity and Basic Control Systems

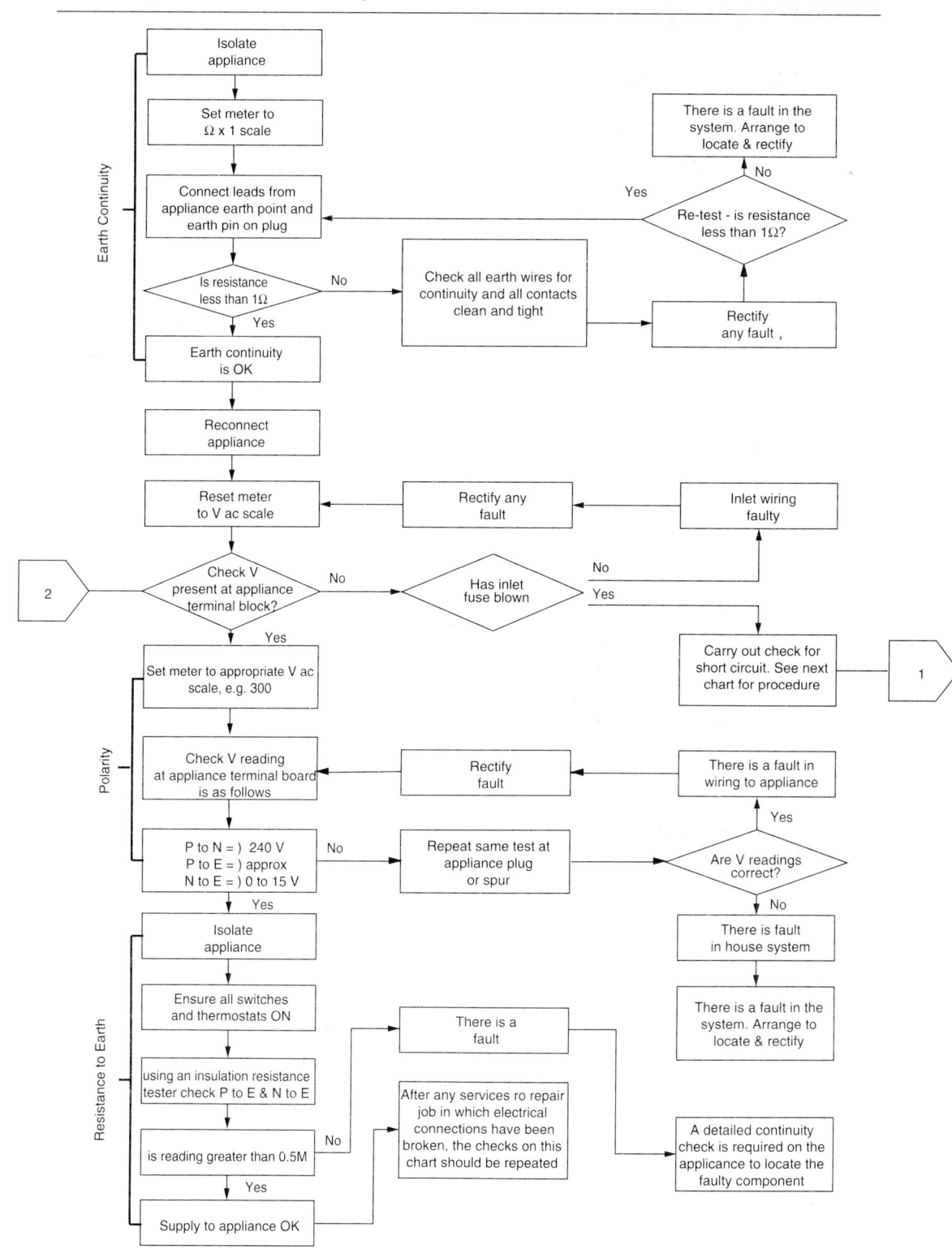

ELECTRICITY SUPPLY CHECK

Earth Continuity

- Isolate appliance
- Set meter to Ω x 1 scale
- Connect leads from appliance earth point and earth pin on plug
- Is resistance less than 1Ω?
  - No → Check all earth wires for continuity and all contacts clean and tight → Rectify any fault → Re-test - is resistance less than 1Ω?
    - No → There is a fault in the system. Arrange to locate & rectify
    - Yes → (back to flow)
  - Yes → Earth continuity is OK
- Reconnect appliance
- Reset meter to V ac scale
- Check V present at appliance terminal block?
  - No → Has inlet fuse blown?
    - No → Inlet wiring faulty → Rectify any fault → (back)
    - Yes → Carry out check for short circuit. See next chart for procedure → **1**
  - Yes ↓

(from **2**) →

Polarity

- Set meter to appropriate V ac scale, e.g. 300
- Check V reading at appliance terminal board is as follows
- P to N =) 240 V
  P to E =) approx
  N to E =) 0 to 15 V
  - No → Repeat same test at appliance plug or spur → Are V readings correct?
    - Yes → There is a fault in wiring to appliance → Rectify fault → (back)
    - No → There is fault in house system → There is a fault in the system. Arrange to locate & rectify
  - Yes ↓

Resistance to Earth

- Isolate appliance
- Ensure all switches and thermostats ON
- using an insulation resistance tester check P to E & N to E
- is reading greater than 0.5M
  - No → There is a fault → After any services ro repair job in which electrical connections have been broken, the checks on this chart should be repeated → A detailed continuity check is required on the appliance to locate the faulty component
  - Yes → Supply to appliance OK

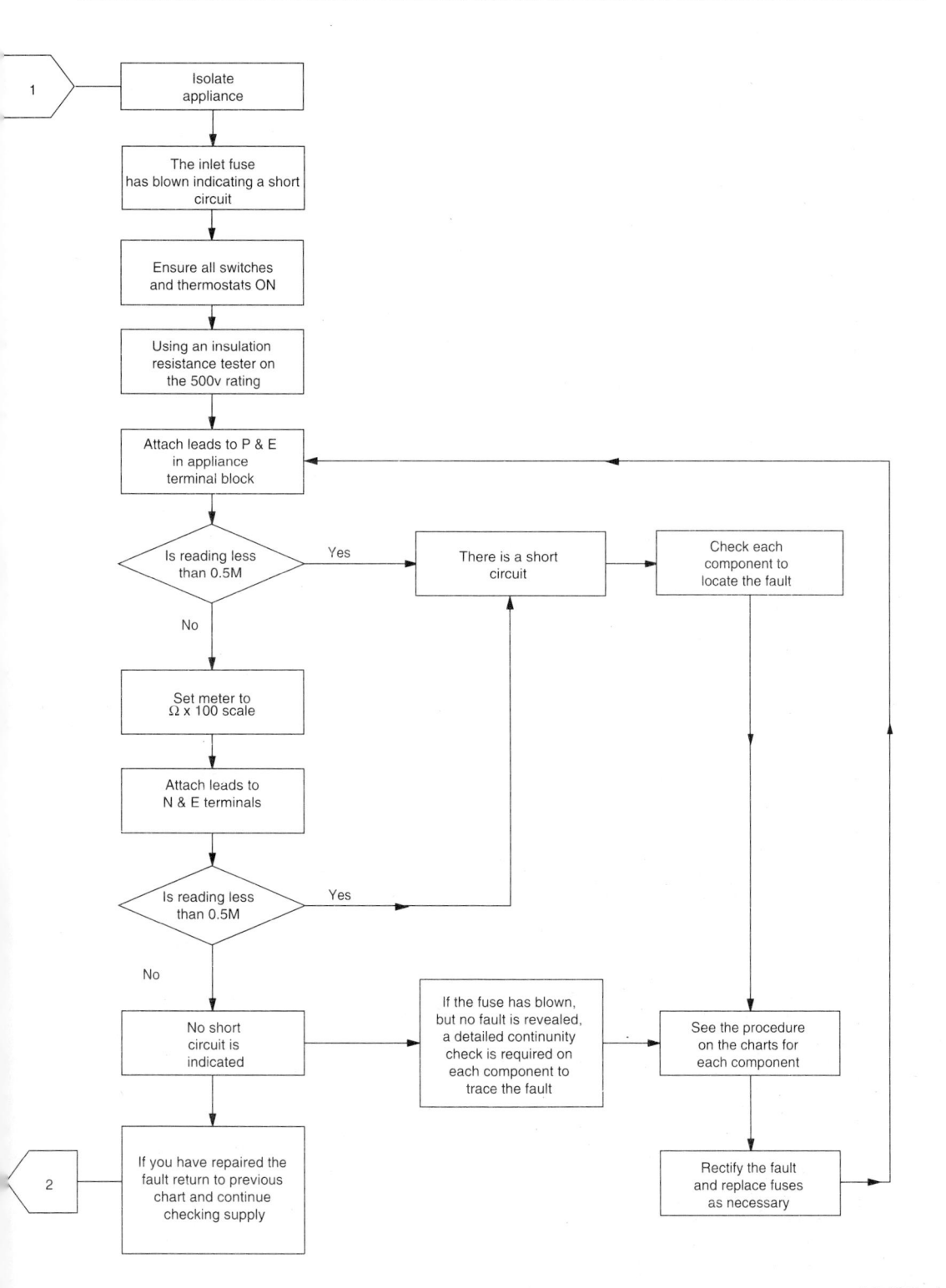

Electricity and Basic Control Systems **CHECK FOR SHORT CIRCUIT**

```
1 ──▶ Isolate appliance
        │
        ▼
   The inlet fuse has blown indicating a short circuit
        │
        ▼
   Ensure all switches and thermostats ON
        │
        ▼
   Using an insulation resistance tester on the 500v rating
        │
        ▼
   Attach leads to P & E in appliance terminal block
        │
        ▼
   Is reading less than 0.5M ──Yes──▶ There is a short circuit ──▶ Check each component to locate the fault
        │ No
        ▼
   Set meter to Ω x 100 scale
        │
        ▼
   Attach leads to N & E terminals
        │
        ▼
   Is reading less than 0.5M ──Yes──▶ (to "There is a short circuit")
        │ No
        ▼
   No short circuit indicated ──▶ If the fuse has blown, but no fault is revealed, a detailed continuity check is required on each component to trace the fault ──▶ See the procedure on the charts for each component
        │
        ▼
   If you have repaired the fault return to previous chart and continue checking supply ──▶ 2

   Rectify the fault and replace fuses as necessary
```

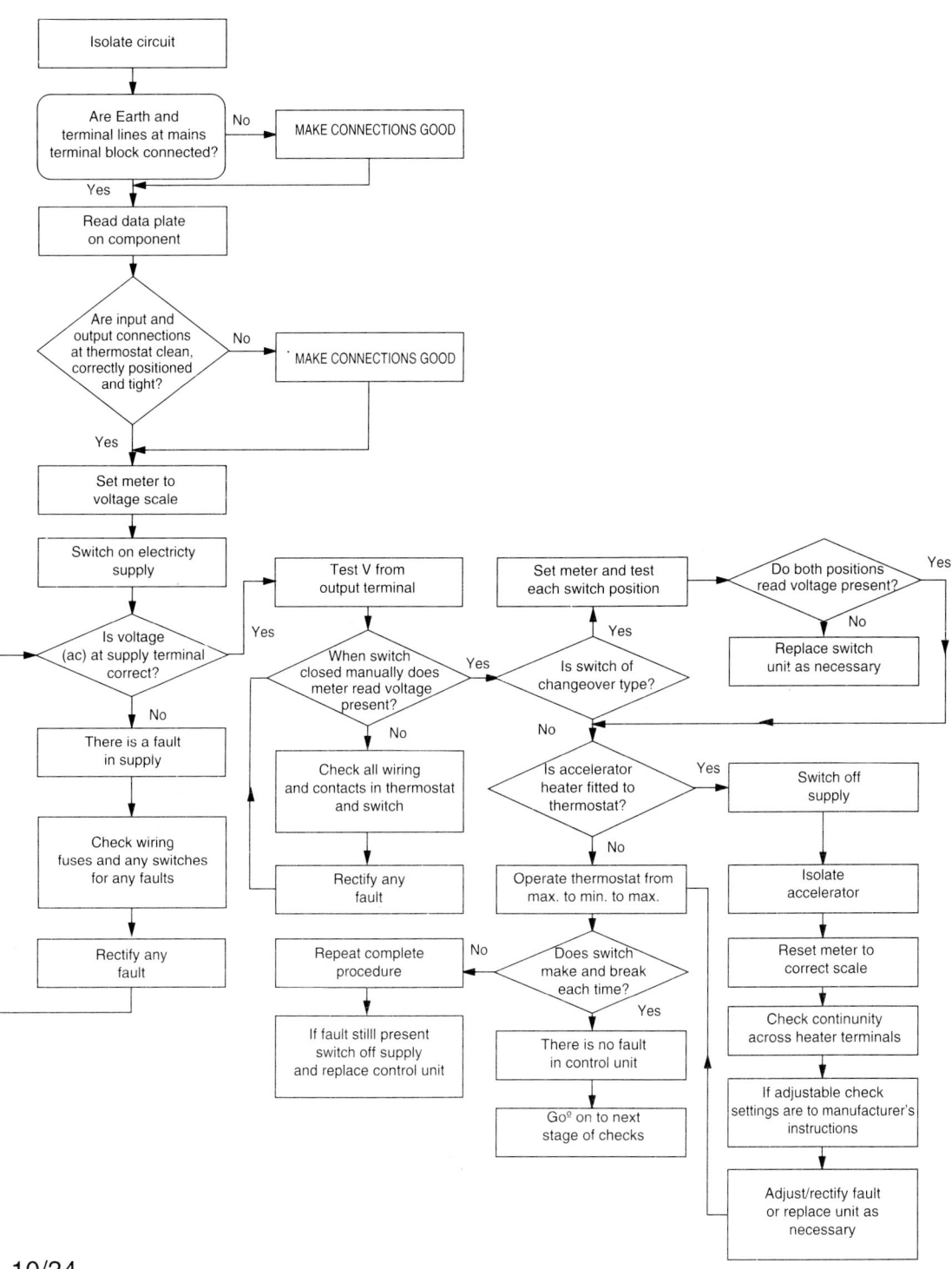

Electricity and Basic Control Systems — THERMOSTATS AND SWITCHES

```
                        Isolate circuit
                              │
                              ▼
              ┌─────────────────────────┐     No
              │ Are Earth and terminal  │─────────► MAKE CONNECTIONS GOOD
              │ lines at mains terminal │
              │   block connected?      │
              └─────────────────────────┘
                         │ Yes
                         ▼
                  Read data plate
                   on component
                         │
                         ▼
              ┌─────────────────────────┐     No
              │  Are input and output   │─────────► MAKE CONNECTIONS GOOD
              │ connections at thermostat│
              │ clean, correctly        │
              │ positioned and tight?   │
              └─────────────────────────┘
                         │ Yes
                         ▼
                   Set meter to
                  voltage scale
                         │
                         ▼
                Switch on electricity          Test V from              Set meter and test          Do both positions      Yes
                     supply                  output terminal            each switch position        read voltage present? ─────┐
                         │                         │                           ▲                           │                   │
                         ▼                   Yes   │                           │ Yes                       │ No                │
              ┌──────────────────┐   Yes   ┌──────────────────┐         ┌──────────────────┐        ┌──────────────────┐       │
              │  Is voltage (ac) │────────►│ When switch      │─Yes────►│  Is switch of    │        │  Replace switch  │       │
              │  at supply       │         │ closed manually  │         │ changeover type? │        │ unit as necessary│       │
              │ terminal correct?│         │ does meter read  │         └──────────────────┘        └──────────────────┘       │
              └──────────────────┘         │ voltage present? │                │ No                                            │
                         │ No              └──────────────────┘                ▼                                               │
                         ▼                         │ No              ┌──────────────────┐    Yes                               │
                There is a fault                   ▼                 │  Is accelerator  │──────────► Switch off               │
                  in supply                 Check all wiring         │  heater fitted to│            supply                    │
                         │                  and contacts in         │   thermostat?    │               │                      │
                         ▼                  thermostat and           └──────────────────┘               ▼                      │
                 Check wiring                  switch                        │ No                  Isolate                     │
               fuses and any                      │                          ▼                   accelerator                   │
               switches for                       ▼                Operate thermostat from           │                         │
                any faults                  Rectify any            max. to min. to max.              ▼                         │
                         │                     fault                         │                  Reset meter to                 │
                         ▼                        │                          ▼                   correct scale                 │
                  Rectify any              Repeat complete    No    ┌──────────────────┐             │                         │
                     fault                  procedure ◄─────────────│  Does switch     │             ▼                         │
                                                 │                  │  make and break  │      Check continuity                 │
                                                 ▼                  │  each time?      │      across heater terminals          │
                                         If fault still present     └──────────────────┘             │                         │
                                         switch off supply and              │ Yes                    ▼                         │
                                         replace control unit               ▼                 If adjustable check              │
                                                                   There is no fault           settings are to manufacturer's  │
                                                                    in control unit              instructions                  │
                                                                            │                         │                        │
                                                                            ▼                         ▼                        │
                                                                     Go on to next           Adjust/rectify fault              │
                                                                    stage of checks          or replace unit as ◄──────────────┘
                                                                                                 necessary
```

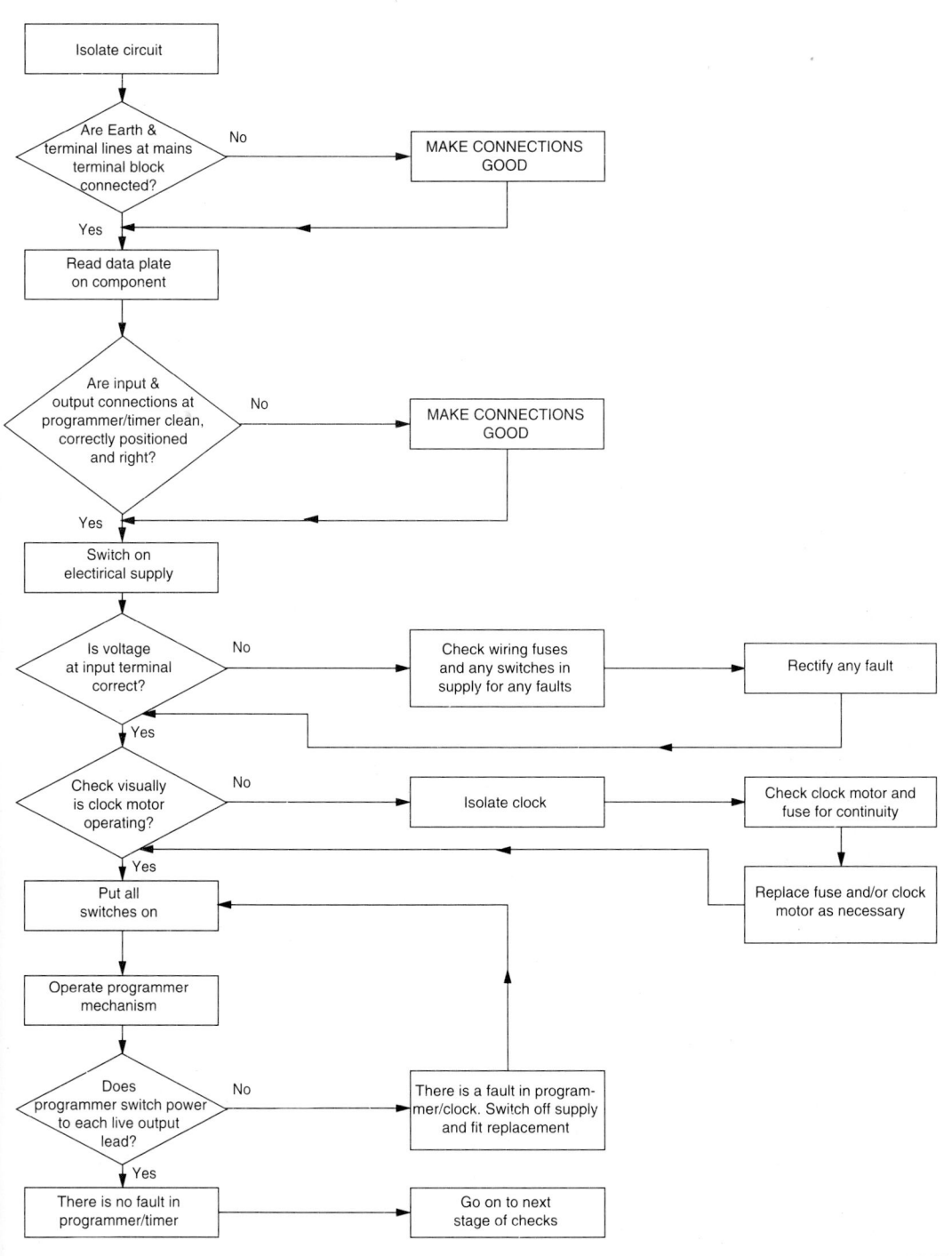

Electricity and Basic Control Systems — PROGRAMMERS AND TIMERS

```
┌──────────────────┐
│ Isolate circuit  │
└────────┬─────────┘
         │
      ╱Are Earth &╲     No
     ╱ terminal lines ╲──────────► ┌─────────────────┐
     ╲ at mains        ╱           │ MAKE CONNECTIONS│
     ╲ terminal block ╱            │ GOOD            │
      ╲ connected?  ╱              └────────┬────────┘
         │Yes                               │
┌──────────────────┐                        │
│ Read data plate  │◄───────────────────────┘
│ on component     │
└────────┬─────────┘
         │
      ╱Are input &  ╲    No
     ╱ output connections╲──────► ┌─────────────────┐
     ╲ at programmer/timer╱       │ MAKE CONNECTIONS│
     ╲ clean, correctly  ╱        │ GOOD            │
     ╲ positioned and right?╱     └────────┬────────┘
         │Yes                              │
┌──────────────────┐                       │
│ Switch on        │◄──────────────────────┘
│ electrical supply│
└────────┬─────────┘
         │
      ╱Is voltage ╲   No    ┌──────────────────┐        ┌──────────────────┐
     ╱ at input    ╲───────►│ Check wiring fuses│──────►│ Rectify any fault│
     ╲ terminal    ╱        │ and any switches  │       └──────────────────┘
     ╲ correct?   ╱         │ in supply for     │
         │Yes               │ any faults        │
         │                  └─────────┬─────────┘
         │◄───────────────────────────┘
      ╱Check visually╲  No   ┌──────────────┐        ┌──────────────────────┐
     ╱ is clock motor╲─────► │ Isolate clock│──────► │ Check clock motor and│
     ╲ operating?   ╱        └──────────────┘        │ fuse for continuity  │
         │Yes                                        └──────────┬───────────┘
         │                                                      │
         │                                          ┌───────────▼──────────┐
         │◄─────────────────────────────────────────│ Replace fuse and/or  │
         │                                          │ clock motor as       │
         │                                          │ necessary            │
┌──────────────────┐                                └──────────────────────┘
│ Put all          │
│ switches on      │
└────────┬─────────┘
         │
┌──────────────────┐
│ Operate programmer│
│ mechanism        │
└────────┬─────────┘
         │
      ╱Does programmer╲  No   ┌──────────────────────┐
     ╱ switch power to ╲─────►│ There is a fault in  │
     ╲ each live output╱      │ programmer/clock.    │
     ╲ lead?          ╱       │ Switch off supply    │
         │Yes                 │ and fit replacement  │
┌──────────────────┐          └──────────────────────┘
│ There is no fault │         ┌──────────────────────┐
│ in programmer/    │────────►│ Go on to next        │
│ timer            │          │ stage of checks      │
└──────────────────┘          └──────────────────────┘
```

Electricity and Basic Control Systems

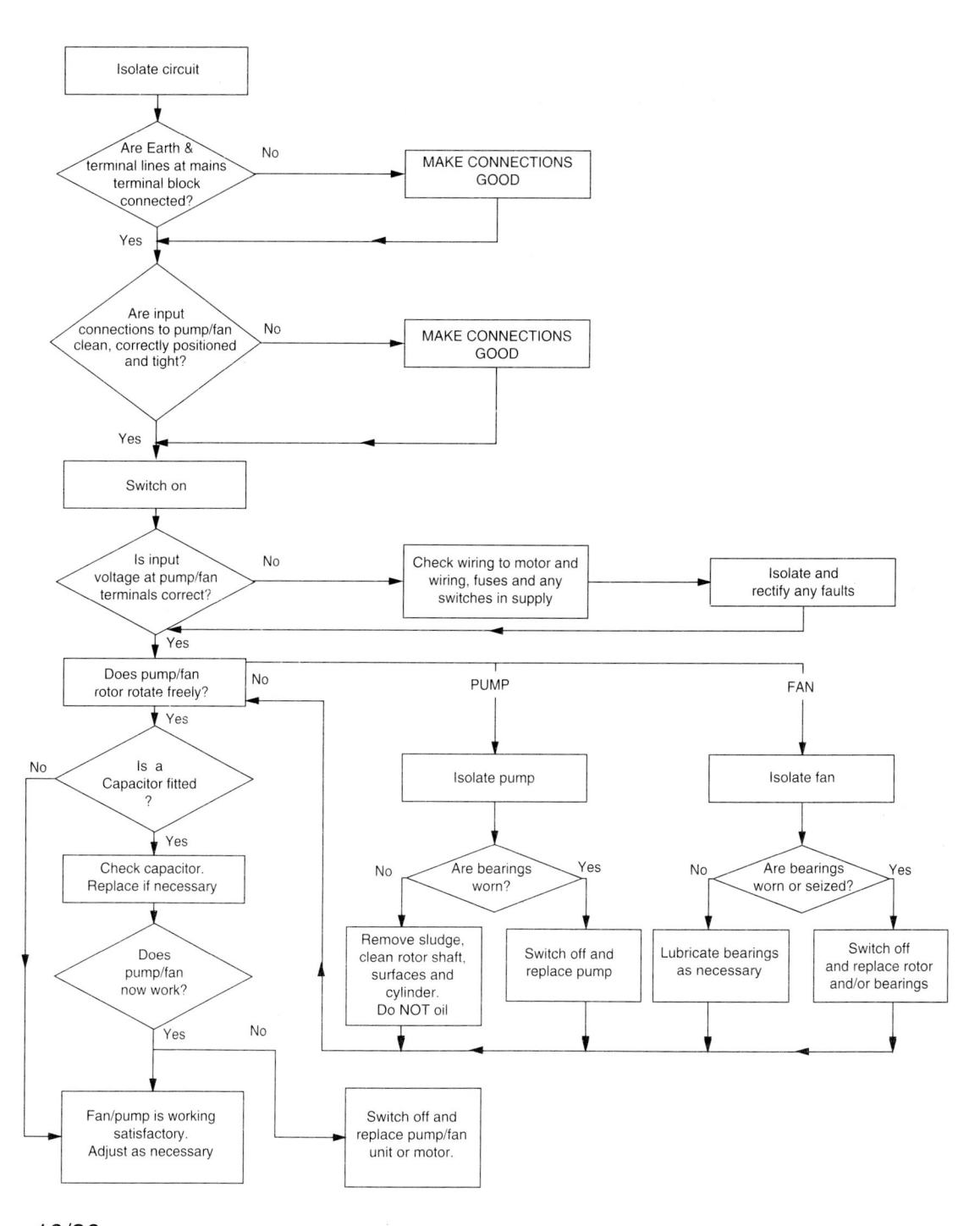

PUMPS AND FANS

```
Isolate circuit
      │
      ▼
Are Earth & terminal lines at mains terminal block connected? ──No──► MAKE CONNECTIONS GOOD
      │Yes
      ▼
Are input connections to pump/fan clean, correctly positioned and tight? ──No──► MAKE CONNECTIONS GOOD
      │Yes
      ▼
Switch on
      │
      ▼
Is input voltage at pump/fan terminals correct? ──No──► Check wiring to motor and wiring, fuses and any switches in supply ──► Isolate and rectify any faults
      │Yes
      ▼
Does pump/fan rotor rotate freely? ──No──► [PUMP: Isolate pump → Are bearings worn? ──Yes──► Switch off and replace pump; ──No──► Remove sludge, clean rotor shaft, surfaces and cylinder. Do NOT oil]
      │Yes                                  [FAN: Isolate fan → Are bearings worn or seized? ──Yes──► Switch off and replace rotor and/or bearings; ──No──► Lubricate bearings as necessary]
      ▼
Is a Capacitor fitted? ──No──► (to next step)
      │Yes
      ▼
Check capacitor. Replace if necessary
      │
      ▼
Does pump/fan now work? ──No──► Switch off and replace pump/fan unit or motor.
      │Yes
      ▼
Fan/pump is working satisfactory. Adjust as necessary
```

Circuit Design

Introduction

Before installing any circuit it is necessary to correctly design it, to comply with the requirements of the IEE Wiring Regulations, so as to ensure the circuit can be used without danger.

The requirements of the IEE Regulations are that:

- the cable used to supply the load should be able to carry the load safely, after taking into account diversity, without cable overheating.
- protective devices such as fuses and MCBs should be able to provide protection for the cable from excess current.
- the length of cable used is limited in length to ensure adequate voltage is provided at the load e.g. shower, to enable it to function safely.
- the maximum value of impedance (resistance) of a circuit from the supply transformer to the load is not exceeded in order to ensure the circuit, under fault conditions, disconnects in 5 seconds for fixed equipment and 0.4 seconds for socket outlet circuits or any circuit in a bathroom.

Maximum Demand and Diversity

Diversity

Consider a domestic installation. It is extremely unlikely that all appliances and equipment will be in full use at any one time; for example, in normal circumstances a householder would be unlikely to switch on all the appliances - kettle, fires, water heaters, iron, toaster and cooker - at the same time, and it would be uneconomical to provide cables and switchgear of a capacity for the maximum possible load; the loads they will carry are likely to be less than the maximum. It is this factor which is referred to as 'diversity'. By making allowances for diversity the size and cost of conductors, protective devices and switchgear can be reduced.

To calculate the diversity factor (formula) for every type of electrical installation, specialist knowledge and experience is required.

The Appendices to 16th Edition of the IEE Regulations do not provide information on this subject; the following information is therefore based on that given in Tables 1A and 1B of The IEE On-Site Guide.

The common methods of obtaining the current of a circuit is to add together the current demand of all points of utilisation and equipment in a circuit.

Methods of Diversity

For the design of an installation one of the following methods may be used.

Method 1

The current demand of a circuit supplying a number of final circuits can be obtained by adding the current demands of all the equipment supplied by each final circuit of the system and applying the allowances for diversity given in Tables 1A and 1B of the IEE On Site Guide. For a circuit with socket outlets the rated current of the protective device is the current demand of the circuit.

Method 2

The current demand of a circuit determined by a suitably qualified electrical engineer.

Example

Consider a small guest house with 10 bedrooms, 3 bathrooms, lounge, dining room, kitchen and utility room with the following loads connected to 240 volt single phase circuits balanced over a 3 phase supply.

Lighting 3 circuits tungsten lighting. Total 2,860 watts

Power 3 x 30A ring circuits to 13A socket outlets

Water heating	1 x 7kw shower
	2 x 3kw immersion heater thermostatically controlled
Cooking appliances	1 x 3kw cooker
	1 x 10.7kw cooker

Calculations and Answers to Example

		Current Demand (Amperes)	Table 1B (Diversity Factor)	Current Demand allowing for Diversity (Amperes)
Lighting	$\frac{2,860}{240}$	11.92	75%	8.94
Power	(i)	30	100%	60
	(ii)	30	50%	
	(iii)	30	50%	
Water Heaters (inst)	$\frac{7,000}{240}$	29.2	100%	29.22
Water Heaters (thermo)	$\frac{6,000}{240}$	25	100%	25
Cookers	(i) $\frac{10,700}{240}$	44.58	100%	44.58
	(i) $\frac{3,000}{240}$	12.5	80%	10

Total Current Demand (allowing Diversity) = 177.72

Load - spread over 3 phases = $\frac{177.72}{3}$

= 59.24A

= 60A per chase

Overload Current

Overload currents usually occur because the equipment is overloaded, (drawing excess current), or the installation is abused, or has been badly designed, or has been modified by an incompetent person. The danger in all such cases is that the temperature of the conductors will increase to such an extent that the effectiveness of any insulating materials will be impaired.

The devices used to detect such overload currents, and to break the circuit when they occur are the FUSE, (either the rewirable or the HBC fuse), and the MINIATURE CIRCUIT BREAKER (MCB).

In order to protect against overload current the protective devices must be rated greater than, or at least equal to the design current; and the current carrying capacity of the cables must be greater than (or equal to) the rating of the protective device.

Protection Against Short-Circuit Currents

Whereas overload currents are likely to result in a current of perhaps no more than twice or three times the normal circuit current, a short-circuit current may be several hundred, or even several thousand times normal. In these circumstances the circuit protection must break the fault current rapidly, before danger is caused through overheating or mechanical stress.

The IEE Regulations stipulate maximum permitted disconnection times of 0.4 seconds for final circuits supplying socket outlets and 5 seconds for circuits to fixed equipment; except where such equipment is situated in a room with a bath or shower when the disconnection time must also be 0.4 seconds.
The current likely to flow under short circuit conditions is called the prospective short-circuit current, the value of which can be measured using a PSC test meter (where the circuit has been installed) or obtained from the supply company.

If the device used for overload protection is also capable of breaking prospective short-circuit current safely, it may be used for both overload and short-circuit protection.

Protective Devices - Fuses

A 'fuse' is simply a short length of wire, chosen for its physical characteristics which 'melts' at a pre-determined temperature, corresponding to a specific electrical current. When the element ' breaks', the circuit is broken and the flow of current is interrupted.

BS 3036

Semi-enclosed or rewirable fuses are now less common. Their main advantages are that they are cheap and easy to repair, but they deteriorate with age and it is always possible for someone to effect a repair with the incorrect size of fuse wire, which may lead to the circuit being overloaded with subsequent risk of damage.

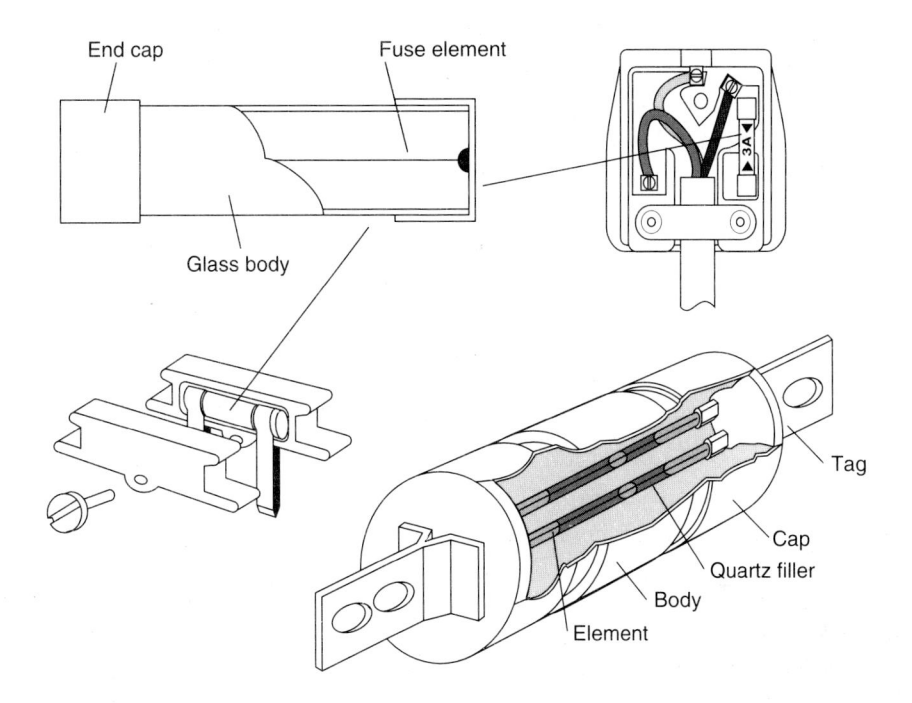

BS 1361 and 1362

Cartridge fuses may be either of ceramic (low grade) or glass with metal end caps, and are sometimes filled with silica sand. Their advantage is their convenience and small physical size. They are manufactured to an accurate current rating and are unlikely to deteriorate. BS 1361 fuses are found in consumer units and BS 1362 fuses in 13 amp plug tops.

BS 88

HBC (High Breaking Capacity) fuses are made from high grade ceramic material to withstand the stress imposed by heavy current interruption. They are sometimes fitted with an indicator (bead) which shows when it is blown. These fuses are found in distribution boards for industrial and commercial premises.

Protective Devices - Miniature Circuit Breakers (MCB)

Miniature circuit breakers (MCB's) are another method of interrupting the flow of current in the event of a fault or overload. They are usually an electro-magnetic device which may include a bi-metallic strip or other temperature sensitive element. MCB's are manufactured to precise specifications and cannot be altered in use.

Their main disadvantage is that they are expensive.

Miniature circuit breakers should be manufactured to BS 3871 and are available in the following types:

Table 2 BS 3871 Instantaneous Tripping Currents of MCB's

Type	Ampere			
1	>	$2.7I_n$	≤	$4.0I_n$
2	>	$4.0I_n$	≤	$7.0I_n$
3	>	$7.0I_n$	≤	$10I_n$
4	>	$10I_n$	≤	$50I_n$
B	>	$3I_n$	≤	$5I_n$
C	>	$5I_n$	≤	$10I_n$
D	>	$10I_n$	≤	$20I_n$

Types 1 and 2 are the types of MCB usually found protecting final circuits in consumer units. For example a Type 2 15 amp MCB should be required to operate within the fault current range 4 x 15 amp (60 amp) to 7 x 15 amp (105 amp). Type 3 MCBs are usually used where motors are connected e.g. compressors or refrigeration units.

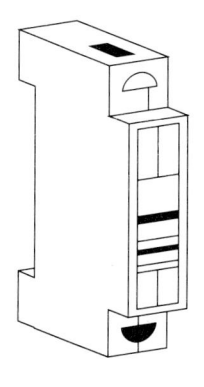

In order to ensure an MCB is capable of handling the prospective short-circuit current of an installation the PSC should be measured using a PSC tester or the value obtained from the electricity company then an MCB with an 'M' category of duty greater than the PSC value should be chosen from the table below:

Category of duty	Prospective current of the test circuit (A)	Power factor of the test circuit
M.1	1000	0.85 to 0.9
M.1.5	1500	0.8 to 0.85
M.2	2000	0.75 to 0.8
M.3	3000	0.75 to 0.8
M.4	4000	0.75 to 0.8
M.6	6000	0.75 to 0.8
M.9	9000	0.55 to 0.6

Protection Against Shock Risk

For conventional final circuits, the IEE Regulations require that under fault conditions that the following disconnection times are not exceeded, in order to minimise the risk of electric shock.

For circuits supplying fixed equipment	5 seconds
For circuits supplying socket outlets	0.4 seconds
For any circuit in a bathroom	0.4. seconds
For any circuit supplying equipment outside the building	0.4 seconds

In order to achieve these disconnection times the IEE Regulations list the maximum impedance values for the different types of protective device for a particular disconnection time or circuit. These tables are 41B1, 41B2 and 41D of the 16th Edition IEE Regulations and Tables 2A, 2B and 2C of the IEE On-Site Guide.

To be able to understand the meaning of maximum loop impedance of a system we should consider the illustration below.

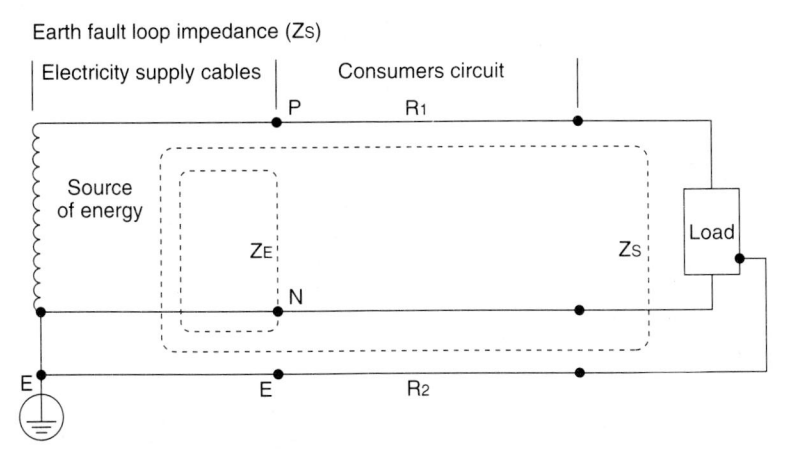

The earth fault loop impedance (Zs) is made up of the impedance of the consumer's phase and protective conductors R1 and R2 respectively, and the impedance external to the installation Z_E (i.e. impedance of the supply). As the value of Z_E will be obtained from the electricity company for the initial assessment of the installation, the maximum impedance allowed for the phase and protective conductors can be determined from:

$$Z_S = Z_E + R_1 + R_2$$

It can therefore be seen that if we obtain the value of external loop impedance for a system from the electricity company then the resistance of the cable for the installation can be obtained e.g. $(R1 + R2) = Z_S - Z_E$. Table 6A of the IEE On Site Guide gives (R1 + R2) resistance values for phase conductors R1 and circuit protective conductors R2. Therefore by dividing the (R1 + R2) values multiplied by the multiplier in Table 6A the maximum length of cable can be established.

Example

A 240V, 20A socket outlet circuit is to be protected by a BS fuse and is wired in $2.5mm^2$. PVC sheathed cable with a $1.5mm^2$ protective conductor. If the value of Z_E is 1.0 ohm determine the maximum length of run. Assume ambient temperature 20°C.

Step 1

Find (Zs) max from Table 41B1(b) 16th Edition IEE Regulations

= 1.78 Ω

Step 2

Find maximum value of (R1 + R2)

= $Z_S - Z_E$
= 1.78 - 1.0
= 0.78 Ω

Step 3

Find (R1 + R2) milli-ohms/metre value from Table 9A IEE On-Site Guide
= 19.51 milli-ohms/metre

Step 4

Apply multiplier from Table 9B IEE On-Site Guide

$$= \frac{19.51 \times 1.38}{1000}$$

$$= 0.027 \ \Omega/m$$

Step 5

Find maximum length of cable run for compliance

$$\text{Max length of cable} = \frac{\text{max allowable } (R_1 + R_2)}{\text{actual } (R_1 + R_2) \ \Omega/m} = \frac{0.78 \ \Omega}{0.027 \ \Omega} = 28.9 \ m\Omega$$

Voltage Drop (525)

A further consideration when selecting cables is that of volt-drop. The IEE Regulations require that under normal service conditions the voltage at the terminals of any fixed current using equipment shall be greater than the lower limit required by the British Standard relevant to the equipment. Where the equipment is not the subject of a British Standard, the voltage at the terminals must be such as not to impair the safe functioning of the equipment.

The requirements of Regulations are satisfied if on a supply given in accordance with the Electricity Supply Regulations 1988 the volt-drop between the origin of the installation (usually the supply terminals) and the fixed current using equipment does not exceed 4% of the voltage.

Current-carrying capacity tables in appendix 4 of the 16th Edition IEE Regulations and appendix 7 of the IEE On-Site Guide, also include values from which a cable volt drop can be calculated for each size of every cable type; the value given is expressed as a voltage drop per ampere per metre of cable (millivolts). To calculate the voltage drop, this figure must be multiplied by the length of the cable (in metres) and the design current on full load. The final product must be divided by 1000 to give the answer in volts.

Diversity must be taken into account when calculating volt drop. The application of rating factors means that in many cases the actual current is much less than the rated current, and the cable is cooler, and this has a lower resistance than that calculated.

Example

A 240 volt single-phase final circuit for a 3kw immersion heater consists of a 22m length of PVC insulated and sheathed cable clipped to the surface. The circuit has a design current of 12.5A. Determine if the chosen 2.5mm^2. cables with a 1.5mm^2 CPC will meet the volt-drop requirements.

Step 1

Maximum voltage drop allowed = 240 x 4 divided by 100 = 9.6V.

Step 2

Actual voltage drop =
mV/A/m x Design Current x Length in Metres divided by 1000.

Step 3

By referring to Table 4D2B 16th Edition IEE Wiring Regulations and Table 7B2 IEE On-Site Guide.
A 2.5mm^2. cable has a voltage drop of 18 mV/A/m.

Step 4

Actual voltage drop =
mV/A/m Design Current x Length divided by 1000
= 18 x 12.5 x 22 divided by 1000
= 4.95 volts

Therefore since the actual voltage drop of 4.95 volts is less than the maximum allowable voltage drop of 9.6 volts it can be seen that the chosen cable is satisfactory.

Circuit Design for Conventional Circuits

The method of correctly designing a circuit to comply with the IEE Regulations involves:

- calculating the design current

- selecting the type and rating of the protective device

- determining and apply any correction factors to the protective device rating
  e.g. temperature factors

- selecting cable from tables

- calculating voltage drop permissible and checking for compliance

- checking circuit complies with shock protection requirements

All these steps involve extensive use of tables and calculation which can get quite involved. The IEE On-Site Guide contains Table 7.1 for conventional circuits which enables the maximum cable run to be established to comply with the IEE Regulations without calculation.

In order to use the table safely the proposed circuit should comply with the following assumptions.

The installation is supplied by one of the following systems:

- TN-C-S with a maximum loop impedance Z_E of 0.35 ohms

- TN-S with a maximum loop impedance Z_E of 0.8 ohms

The final circuit is to be connected to a consumer unit at the origin of the installation.

The installation method used complied with those specified in the 16th Edition IEE Regulations Table 4A for:

- Method 1 sheathed cables clipped or embedded in plaster

- Method 3 cables run in conduct or trunking

- Method 4 sheathed cables or cables in conduct, in thermal insulation but in contact with a thermally conductive surface on one side only.

Throughout the length of the circuit the temperature does not exceed 30°C.

No grouping factors for cables have been allowed for in circuits protected by protective devices in excess of 6 amps.

Using Table 7.1

Example

A 12.5 amp immersion heater circuit protected by a Type 2 15 amp MCB wired in 2.5mm^2 PVC sheathed cable with a 1.5mm^2 CPC is supplied from a 240 volt single phase TN-C-S system. What is the maximum length of run?

Radial Circuit

Current Rating	Cable CPC Size	Protective Device	Max. length in metres			
			TN-S 0.4s	TN-S 5s	TN-C-S 0.4s	TN-C-S 5s
15	2.5 1.5	Type 2 MCB			35	

Circuit Design Example

An 8kw electric shower is to be installed in the bathroom of a domestic dwelling, the length of proposed cable route is 30 metres clipped to the surface supplied from the spare way in a consumers unit fitted with MCBs.

The electricity supply is a 240 volt TN-C-S system. Design a suitable installation to comply with the requirements of the IEE Regulations using Table 7.1 the of IEE On-Site Guide to minimise the number of calculations required.

Step 1

Determine full load current (I)

$$I = \frac{watts}{volts} = \frac{8,000}{240}$$

I = 33.3A

Step 2

Determine design current
No diversity allowance on showers
Design current I = 33.3A.

Step 3

Select type and size of protective device.
Since the consumer unit is fitted with MCB's Type 2 use Type 2 MCB 45 amp since this is the next size of MCB over 30 amp.

Step 4

Apply correction factors e.g. for temperature
None apply

Step 5

Select cable type and size

Since cable is to be clipped to the surface use PVC insulated and sheathed from current rating tables.

4D2A 16th Edition IEE Regulations
7.1 IEE On-Site Guide

Suitable cable 10mm^2 rated at 63 amps clipped to the surface.

Step 6

Check current complies with shock protection requirements of 0.4 seconds since the shower is fixed equipment installed in a bathroom.

Using Table 7.1 IEE On-Site Guide.

For a 10mm^2 cable with a 4mm^2 CPC using a 45A MCB Type 2 on a TN-C-S system.

The maximum length of run is 43 metres.

Since the actual cable run is 30 metres which is less than 43 metres maximum, the circuit will comply with the IEE Regulations.

Summary

Before installing any conventional final circuit the following must be considered:

- can the consumer unit or distribution board carry the planned additional load?
- what type of protective device is to be used?
- what are the ratings of the protective devices available?
- what type of cable and installation method is to be used?
- what type of earthing arrangement is being used?
- what is the maximum disconnection time for the circuit 0.4s or 5s?
- if system a TT Type an RCD to BS 4293 must be installed.
- what isolation and switching requirements are necessary?
- what labels are required to be fitted?
- is the loop impedance value below the maximum allowed in Table 7.2.3 of the IEE On-Site Guide?

Conventional Circuit Design

When a circuit needs designing which does not fit the requirements of Table 7.1 of the IEE On-Site Guide the following design system may be used.

In order to simplify the stages the charts and cable selection procedure sheets should be used.

Procedure for the Selection of a final Circuit Cable and Protective Device

STEP BY STEP PROCEDURE FOR THE SELECTION OF A FINAL CIRCUIT CABLE AND PROTECTIVE DEVICE

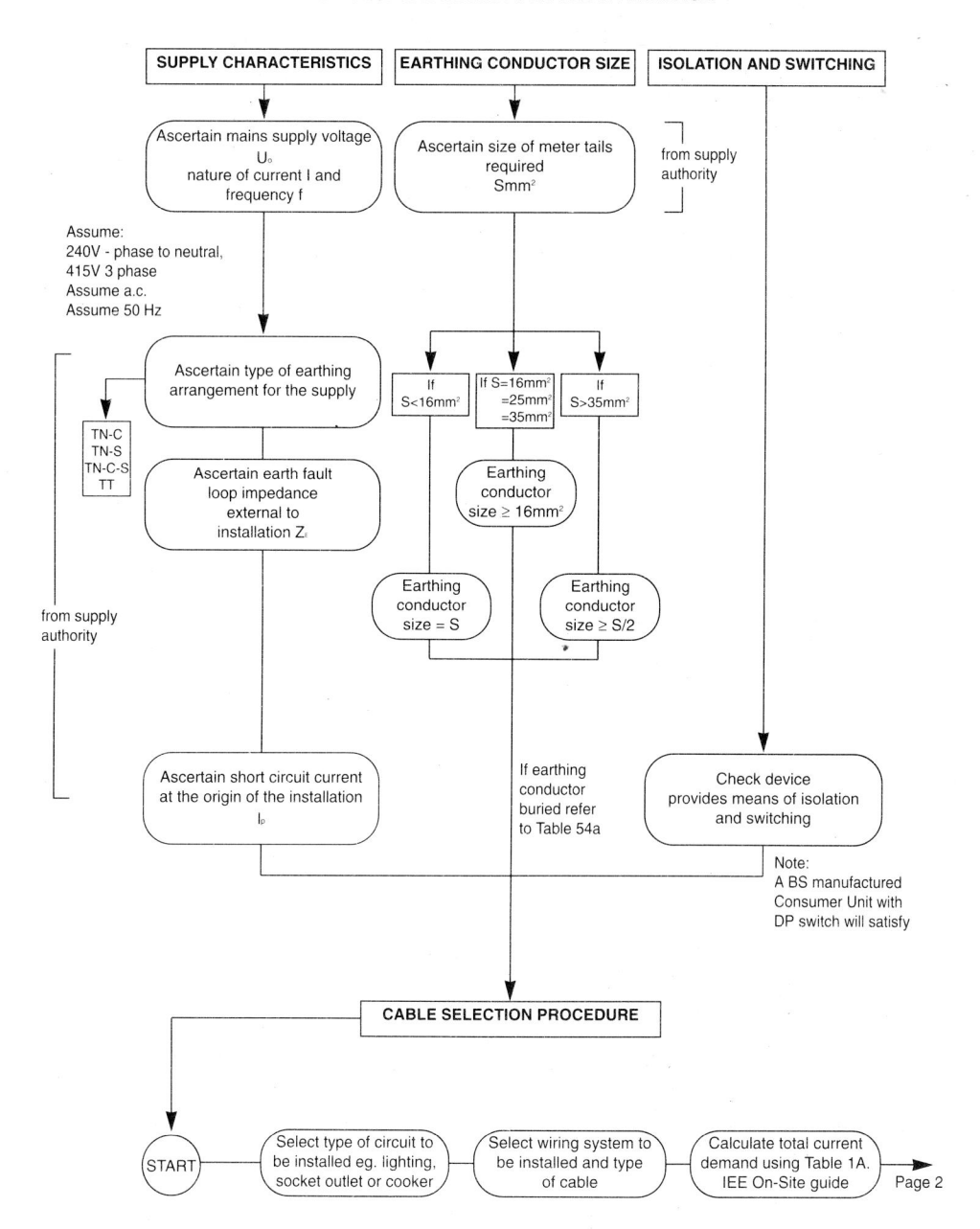

SUPPLY CHARACTERISTICS

Ascertain mains supply voltage U_o nature of current I and frequency f

Assume:
240V - phase to neutral,
415V 3 phase
Assume a.c.
Assume 50 Hz

Ascertain type of earthing arrangement for the supply

- TN-C
- TN-S
- TN-C-S
- TT

Ascertain earth fault loop impedance external to installation Z_e

from supply authority

Ascertain short circuit current at the origin of the installation I_p

EARTHING CONDUCTOR SIZE

Ascertain size of meter tails required $S mm^2$

- If $S<16mm^2$
- If $S=16mm^2$ $=25mm^2$ $=35mm^2$
- If $S>35mm^2$

Earthing conductor size $\geq 16mm^2$

Earthing conductor size = S

Earthing conductor size $\geq S/2$

If earthing conductor buried refer to Table 54a

ISOLATION AND SWITCHING

from supply authority

Check device provides means of isolation and switching

Note:
A BS manufactured Consumer Unit with DP switch will satisfy

CABLE SELECTION PROCEDURE

START → Select type of circuit to be installed eg. lighting, socket outlet or cooker → Select wiring system to be installed and type of cable → Calculate total current demand using Table 1A. IEE On-Site guide → Page 2

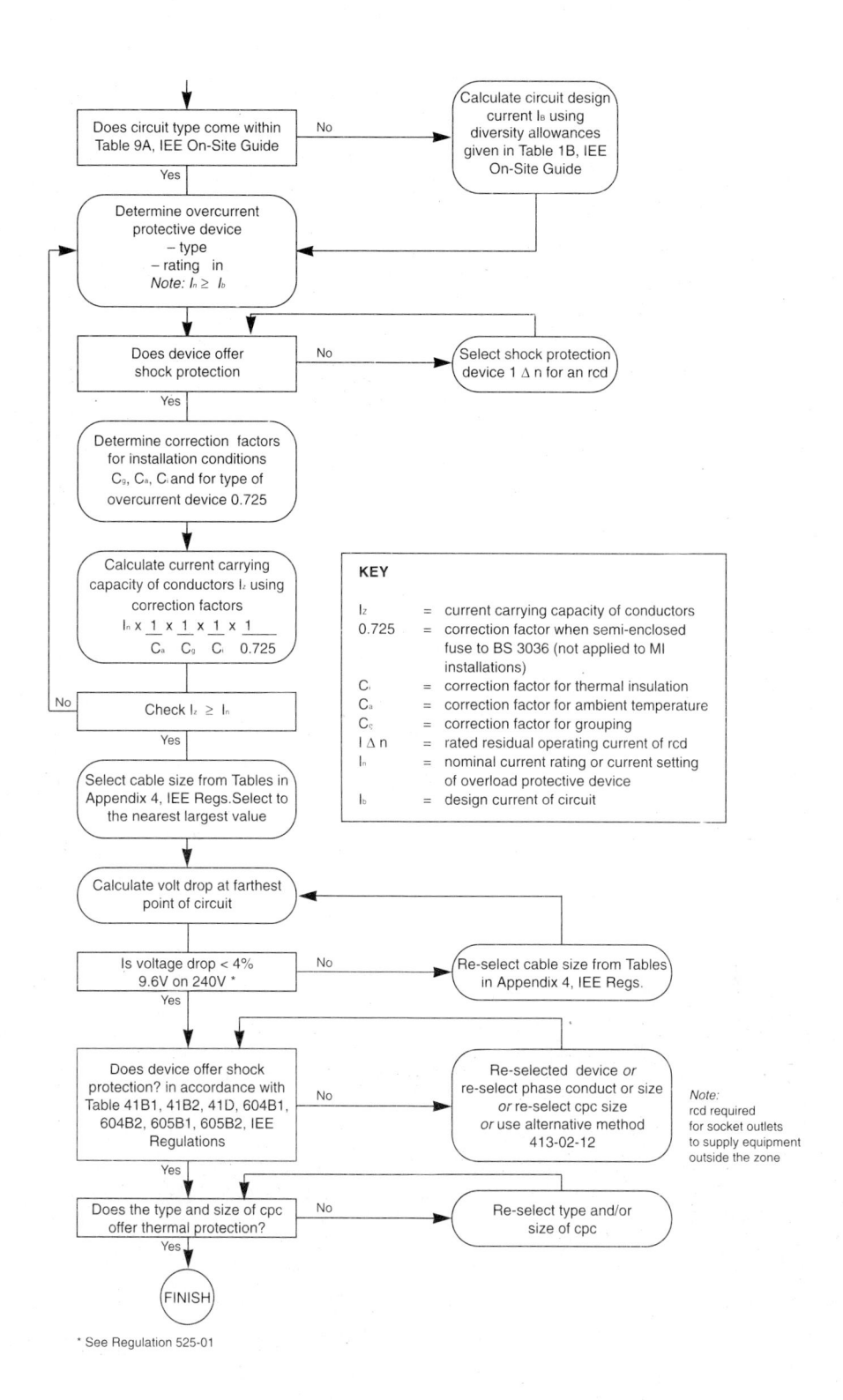

```
                                                  ┌─────────────────────────┐
                ↓                                  │ Calculate circuit design│
   ┌──────────────────────────┐        No         │ current Ib using        │
   │ Does circuit type come within├──────────────→ │ diversity allowances    │
   │ Table 9A, IEE On-Site Guide │                 │ given in Table 1B, IEE  │
   └──────────────────────────┘                    │ On-Site Guide           │
                │ Yes                              └─────────────────────────┘
                ↓                                              │
   ┌──────────────────────────┐                                │
   │ Determine overcurrent    │←───────────────────────────────┘
   │ protective device        │
   │  – type                  │
   │  – rating  in            │
   │ Note: In ≥ Ib            │
   └──────────────────────────┘
```

Flow chart (circuit design procedure):

- Does circuit type come within Table 9A, IEE On-Site Guide?
  - No → Calculate circuit design current I_b using diversity allowances given in Table 1B, IEE On-Site Guide
  - Yes → Determine overcurrent protective device – type – rating in Note: $I_n \geq I_b$

- Does device offer shock protection?
  - No → Select shock protection device $1\,\Delta\,n$ for an rcd
  - Yes → Determine correction factors for installation conditions C_g, C_a, C_i and for type of overcurrent device 0.725

- Calculate current carrying capacity of conductors I_z using correction factors
  $$I_n \times \frac{1}{C_a} \times \frac{1}{C_g} \times \frac{1}{C_i} \times \frac{1}{0.725}$$

- Check $I_z \geq I_n$
  - No → (back to Determine overcurrent protective device)
  - Yes → Select cable size from Tables in Appendix 4, IEE Regs. Select to the nearest largest value

- Calculate volt drop at farthest point of circuit

- Is voltage drop < 4% 9.6V on 240V *
  - No → Re-select cable size from Tables in Appendix 4, IEE Regs.
  - Yes → Does device offer shock protection? in accordance with Table 41B1, 41B2, 41D, 604B1, 604B2, 605B1, 605B2, IEE Regulations
    - No → Re-selected device or re-select phase conduct or size or re-select cpc size or use alternative method 413-02-12
    - Yes → Does the type and size of cpc offer thermal protection?
      - No → Re-select type and/or size of cpc
      - Yes → FINISH

KEY

I_z	=	current carrying capacity of conductors
0.725	=	correction factor when semi-enclosed fuse to BS 3036 (not applied to MI installations)
C_i	=	correction factor for thermal insulation
C_a	=	correction factor for ambient temperature
C_g	=	correction factor for grouping
$I\,\Delta\,n$	=	rated residual operating current of rcd
I_n	=	nominal current rating or current setting of overload protective device
I_b	=	design current of circuit

Note: rcd required for socket outlets to supply equipment outside the zone

* See Regulation 525-01

QUANTITY OR FACTOR	SYMBOL	REFERENCE	DATA CALCULATIONS	VALUE
Load current	I	Diversity tables 1A & 1B, IEE On-Site Guide		
Design current	I_b			
Protective device rating	I_n	Tables 41B1, 41B2, 41D, 604B1, 604B2, 605B1, 605B2 or time/current curves. Appx. 3, IEE Regs		
Design current			Check $I_b \leq I_n$	YES/NO
Cable size		See attached notes		
Grouping factor	C_g	Tables 4C1 and 4C2		
Ambient temperature	C_a	Reg. 523-04 and Table 52A, IEE Regs		
Thermal insulator factor	C_i			
BS 3036 fuse factor	0.725	$I_t \geq \dfrac{I_n}{C_a \times C_g \times C_i \times 0.725}$		
Current carrying capacity	I_t			
Cable size chosen				
Tabulated current carrying capacity	I_t	From tables 4D1A-4L4A, IEE Regs		
Effective current carrying capacity	I_z	$I_t \geq I_z$		
			Check $I_b \leq I_n \leq I_z$	YES/NO
Permissable V drop		Safe functioning of equipment (or 4%) Regs. 525-01		
Actual V drop		$\dfrac{mV/Am \times I_b \times length}{1000}$		
			Check actual voltage drop $\leq$ Permissable voltage drop	YES/NO

QUANTITY OR FACTOR	SYMBOL	REFERENCE	DATA CALCULATIONS	VALUE
Shock protection				
Maximum earth fault loop impedance	Max Z_s	Tables 41B1, 41B2, 41D, 604B1, 604B2, 605B1, 605B2 or IEE Regs		
Exteranl earth fault loop impedance	Z_E	Measure or obtain from electricity company		
c.p.c. size chosen		Select from BS 6004 initially		
Resistance of phase conductor	R_1	Table 9A and 9B, IEE On-Site Guide, manufacturers data or BS standards data		
Resistance of c.p.c	R_2			
Actual earth fault loop impedance	Z_s	$Z_s = Z_E + (R_1 + R_2)$	Check $Z_s = \leq$ Max	YES/NO
Thermal constraint				
Fault current	I_f	$I_f = \dfrac{U_o}{Z_s}$ (240V phase to earth)		
Time	t	Read time/current characteristics Tables 54B to F, IEE Regs		
Factor for c.p.c.	k			
Size c.p.c.	S	$\sqrt{\dfrac{I^2 t}{k}}$ mm^2	Check actual c.p.c. used is $\geq$ than above	YES/NO

Cable Selection Procedure

Selecting Cables for Circuits and Checking for Compliance with Regulations

Certain tables and formulae used in this section are not part of the 16th Edition IEE Regulations. Reference has been made to the IEE On-Site Guide.

When installing a circuit, it is necessary to:

- calculate the design current (I_b)
- select the type and nominal rating of the protective device (I_n)
- determine and apply correction factors to I_n
- select cable from tables in Appendix 4, IEE Regulations (I_z)
- calculate the voltage drop and check for compliance
- check that circuit complies with shock protection
- check that circuit complies with thermal constraints

Example 1

A 20A radial socket outlet circuit is protected by a 20 amp Type 2 MCB. The circuit is wired using 2.5mm^2 PVC sheathed cable with a 1.5mm^2 CPC for 16 metres clipped in the surface. Assuming that no rating factors are applicable and that the value of Z_E is given as 0.5 ohms, determine whether the circuit complies with the IEE Regulations. The nominal voltage (U_o) may be taken as 240V.

I_b Design current of circuit = 20A
I_n Protective device, MCB Type 2 = 20A

Step 1
Apply correction factors for grouping and ambient temperature.
None apply.

Step 2
Select suitable cable (I_z) from Appendix 4
from Table 4D2A IEE Regulations, 2.5mm^2 cable = 27A

Step 3
Calculate voltage drop
Maximum volt-drop allowed = 4% of 240 = 9.6V
Actual voltage drop = $\dfrac{mV/A/m \times \text{design current } I_b \times \text{length}}{1000}$

From Table 4D2B   mV/A/m for 2.5mm^2 cable = 18mV/A/m

∴ Actual v.d. = $\dfrac{18 \times 20 \times 16}{1000}$ = 5.76v

Step 4

Check for compliance with shock protection

Total earth fault loop impedance = Z_s

Maximum Z_s from Table 41B2 IEE Regulations = 1.71 ohms

Actual $Z_s = Z_E$ = (R1 + R2 ohms/metre x length)
Z_E = 0.5 ohms

From Table 9A of the IEE On-Site Guide

(2.5mm^2 phase conductor, 1.5mm^2 protective conductor)

R1 + R2 = 19.51 milliohm/m

From Table 9B of the IEE On-Site Guide

Multiplier for PVC = 1.38

(R1 + R2) = 19.51 x 1.38 x length of run
$$= \frac{19.51 \times 1.38 \times 16}{1000} = 0.43 \, \Omega$$

Actual Z_s = 0.5 + 0.43
= 0.93

Z_s satisfactory i.e. 0.93 < 1.71

Step 5

Check for compliance with thermal constraints
Calculate value of fault current If = $\frac{U_o}{Z_s}$

$$If = \frac{240}{0.93}$$

$$= 258A$$

Find value of t when If = 258A from time/current characteristic for Type 2 20A MCB in Appendix 3 Fig. 5 of the IEE Regulations.

t will be less than 0.01 seconds but for the purpose of the calculation use t = 0.01 seconds

Find value of k from Table 54C
k = 115

minimum size of protective conductor $s = \sqrt{\frac{I^2 t}{k}}$ mm^2

$$s = \sqrt{\frac{258^2 \times 0.01}{115}}$$

$$s = 0.22 \text{mm}^2$$

Nearest standard size of conductor = 1.0mm^2. This will satisfy thermal constraints.

Inspection and Testing

Introduction

Detailed methods of testing are no longer contained in the appendices of the 16th Edition. In order to provide a more comprehensive approach to testing, reference should be made to the IEE On-Site Guide and the IEE Guidance Notes No. 3 Inspection and Testing.

General

Inspection should comprise careful scrutiny of the installation, supplemented by testing to:

1. Verify safety of persons and livestock.
2. Verify protection against damage to property by fire and heat.
3. Establish that the installation is not damaged and has not deteriorated.
4. Identify installation defects or non-compliance with regulations.

Note: Attention is drawn to HSE Guidance Note GS38, 'Electrical Test Equipment for Use by Electricians' published by HMSO, which advises on the selection and safe used of suitable test probes, leads, lamps, voltage indicating devices and other measuring equipment.

Initial Inspection

Reasons for Inspection and Testing

The purpose of inspection of electrical installations is to verify that installations are safe and comply with the requirements of Regulations.

Test Instruments

Test instruments should be regularly checked and re-calibrated to ensure accuracy. The serial number of the instrument used should be recorded with test results, to avoid unnecessary re-testing if one of a number of instruments is found to be inaccurate.
For operation, use and care of test instruments, refer to manufacturer's handbook.

General

During its installation or on completion, every installation must be inspected and tested before being connected to the supply and energised. This should be done in such a manner that no danger to persons or damage to property or equipment can occur, even if the circuit tested is defective.

The following information should be made available to the persons carrying out the inspection and testing of an installation.

Diagrams, charts or tables indicating:

    (a) the type of circuits
    (b) the number of points installed
    (c) the number and size of conductor
    (d) the type of wiring system

The location and types of devices used for:

    - protection
    - isolation and switching

Details of the characteristics of the protection devices for automatic disconnection, the earthing arrangements for the installation, the impedances of the circuits and a description of the method used.

Details of circuits or equipment sensitive to tests - e.g. central heating controls with electronic timers and displays

Note: Information may be given in a schedule for simple installations. See example below for a domestic installation. A durable copy of the schedule relating to distribution board must be provided inside or adjacent to the distribution board.

Schedule of Installation at ....

Type of circuit	Points served	Phase Conductor mm^2	Protective Conductor mm^2	Protective devices	Type of wiring
Lighting	10 downstairs	1mm^2	1mm^2	5 Amp Type 2 MCB	PVC/PVC
Lighting	8 upstairs	1mm^2	1mm^2	5 Amp Type 2 MCB	PVC/PVC
Immersion Heater	Landing	2.5mm^2	1.5mm^2	15 Amp Type 2 MCB	PVC/PVC
Ring	10 downstairs	2.5mm^2	1.5mm^2	30 Amp Type 2 MCB	PVC/PVC
Ring	8 upstairs	2.5mm^2	1.5mm^2	30 Amp Type 2 MCB	PVC/PVC
Shower	Bathroom	6mm^2	2.5mm^2	30 Amp Type 2 MCB	PVC/PVC

Inspection

A detailed inspection should be made of installed electrical equipment, usually with the part of the installation being inspected disconnected from the supply. The inspection should verify that it:

- complies with the British Standards or harmonised European Standards (this may be ascertained by mark or by certificate furnished by the installer or manufacturer)
- is correctly selected and erected in accordance with these Regulations
- is not visibly damaged so as to impair safety

The detailed inspection must include the following where relevant:

Connections of conductors

Identification of conductors

Routing of cables in safe zones or mechanical protection methods

Selection of conductors for current-carrying capacity and voltage drop

Connection of single pole devices for protection or switching in phase conductors only

Correct connection of socket outlets and lampholders

Presence of fire barriers and protection against thermal effects

Methods of protection against direct contact (including measurements of distances where appropriate) i.e.

- protection by insulation of live parts
- protection by barriers or enclosures

Methods of protection against indirect contact:

- presence of protective conductors
- presence of earthing conductors
- presence of supplementary equipotential bonding conductors
- earthing arrangements for combined protective and functional purposes
- use of Class II equipment or equivalent insulation - electrical separation

Prevention of mutual detrimental influence

Presence of appropriate devices for isolation and switching

Choice and setting of protective and monitoring devices

Labelling of circuits, fuses, switches and terminals

Selection of equipment and protective measures appropriate to external influences

Presence of undervoltage protective devices

Adequency of access to switchgear and equipment

Presence of danger notices and other warning notices

Presence of diagrams, instructions and similar information

Erection methods

Note: During any re-inspection of an installation all pertinent items in the check list should be covered.

Testing

The following items (where relevant to the installation being tested), must be tested in the following sequence:

Continuity of protective conductors including main and supplementary bonding

Continuity of ring circuit conductors

Insulation resistance

Polarity

Earth loop impedance

Operation of residual current devices

Suitable reference methods of testing are described in the guide notes on the Regulations. The use of other methods of testing is not precluded provided that they will give results which are no less effective.

If a test indicates failure to comply, that test, and the preceding test (whose results may have been affected by the fault) must be repeated after rectification of the fault.

Continuity of Protective Conductors

The initial tests applied to protective conductors are intended to verify that the conductors are both correctly connected and electrically sound, and also the resistance is such that the overall earth fault loop impedance of the circuits is of a suitable value to allow the circuit to be disconnected from the supply in the event of an earth fault, (within the disconnection times selected to meet the requirements of Regulation 413-02-09) 0.4 seconds or 5 seconds.

Every protective conductor, including main bonding conductors and supplementary bonding conductors, should be tested to verify that the conductors are electrically sound and correctly connected.
Use a low resistance ohmmeter for these tests.

Methods 1

Step 1

Connect one terminal of the ohmmeter to a long test lead and connect this to the consumer's earth terminal, as illustrated.

Step 2

Connect the other terminal of the Ohmmeter to a short lead and use this to make contact with the protective conductor at various points on the installation, testing such items as switch boxes and socket outlets.

The resistance reading obtained by the above method actually includes the resistance of the test leads. Therefore the resistance values of the test leads should be measured and this value deducted from any resistance reading obtained for the installation under test.

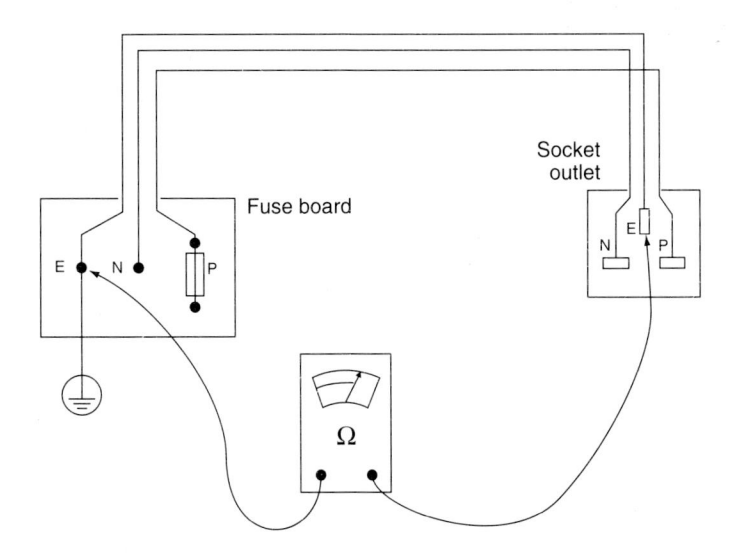

If the distance between the fuseboard and circuit under test involves the use of very long test leads, an alternative (Method 2) using the phase conductors as a test lead may be used.

Strap the phase conductor to the protective conductor at a distant socket outlet, so as to include all of the circuit and test between phase and earth terminals at the fuseboard as illustrated.

The resistance measured by the above method includes the resistance of the phase conductor from the main switch to the point under test.

The approximate resistance of this conductor can be obtained by joining together the phase and neutral conductors at the socket outlet (at the point under test) and measuring the resistance as shown below. The value of conductor resistance is half the value obtained by this test.

The value of circuit protective conductor resistance is calculated as the initial reading, minus phase conductor resistance.

Method 2 (cont'd)

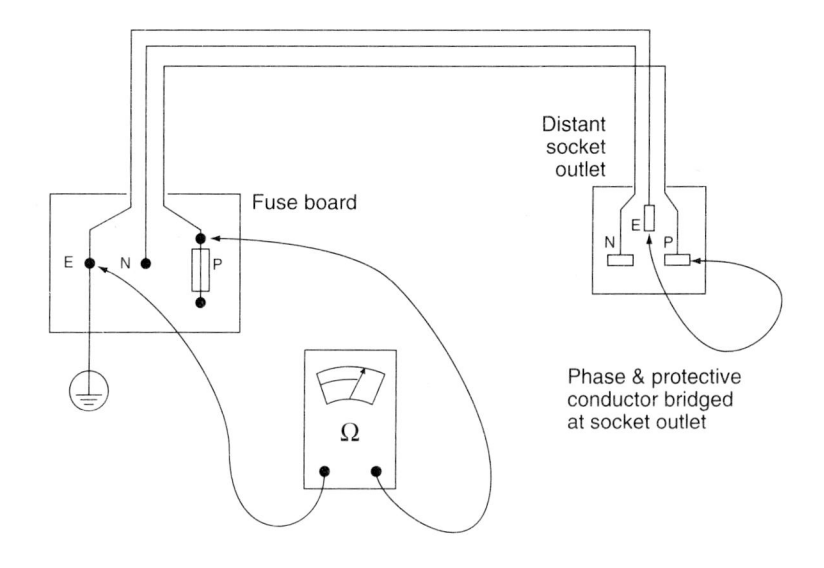

Distant socket outlet

Fuse board

Phase & protective conductor bridged at socket outlet

Alternative to Method 2

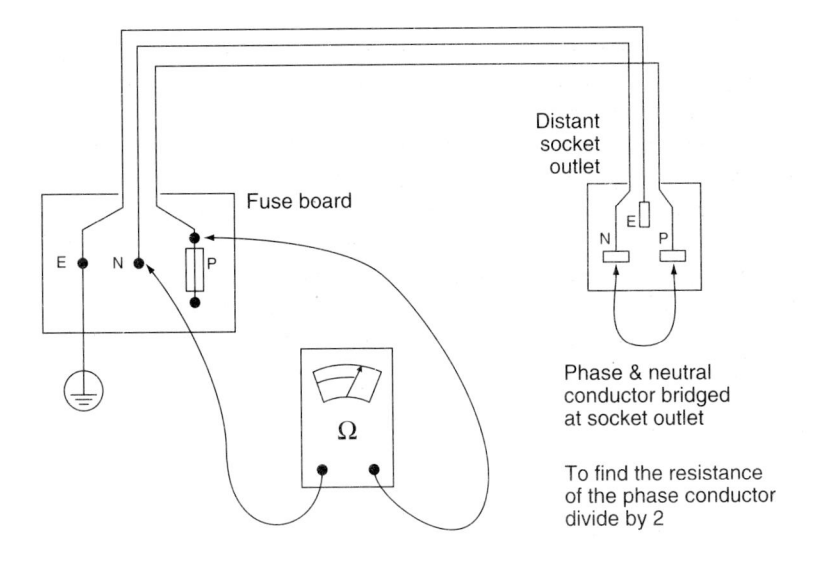

Distant socket outlet

Fuse board

Phase & neutral conductor bridged at socket outlet

To find the resistance of the phase conductor divide by 2

The test methods detailed below, as well as checking the continuity of the protective conductor, also provide a measure of (R1 + R2) which enables the designer to verify the calculated earth fault loop impedance Zs.

A low resistance continuity tester should be used for these tests.

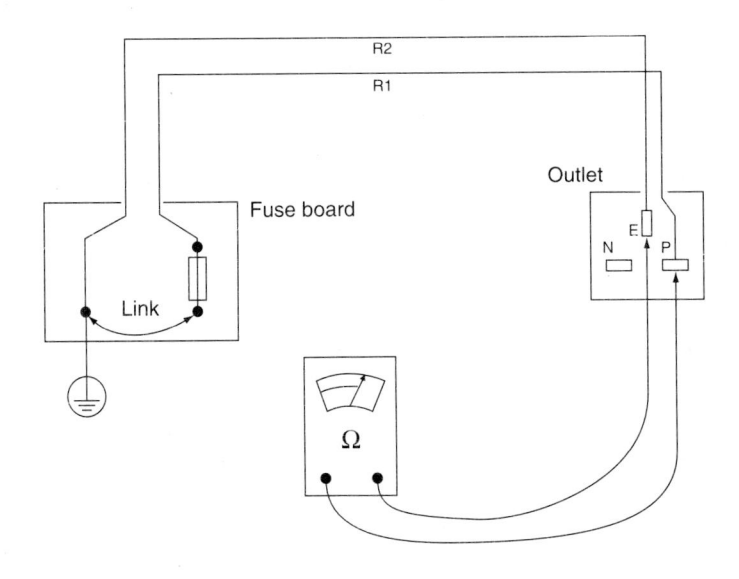

Outlet

Fuse board

Link

Continuity of Ring Final Circuits Conductors (713-03)

A test must be made to verify the continuity of all live and protective conductors to every final ring circuit.

A number of test methods can be used.

The tests are to establish that the ring has not been interconnected to create an apparently continuous ring circuit where an actual break exists, as illustrated below. One core illustrated for clarity.

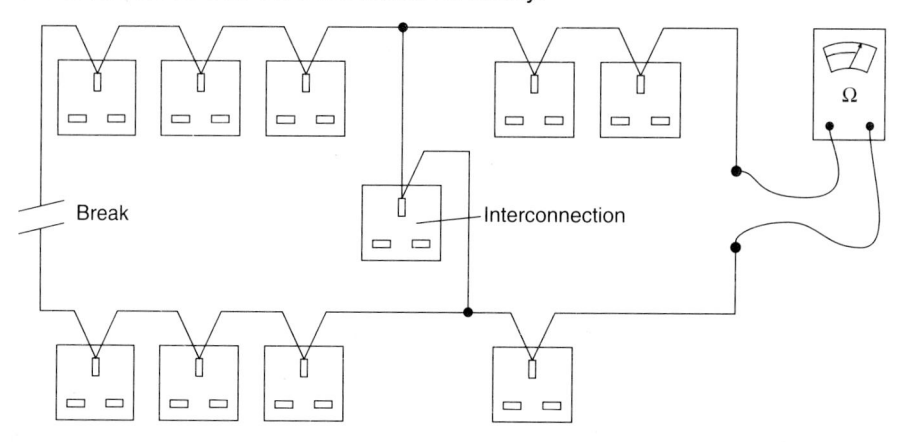

The Test

This test must be made to verify the continuity of phase, neutral and protective conductors of every final ring circuit. The test result should also establish that the ring has not been interconnected to create an apparently continuous ring circuit which is actually broken.

Using a low-resistance ohmmeter, follow the method shown.

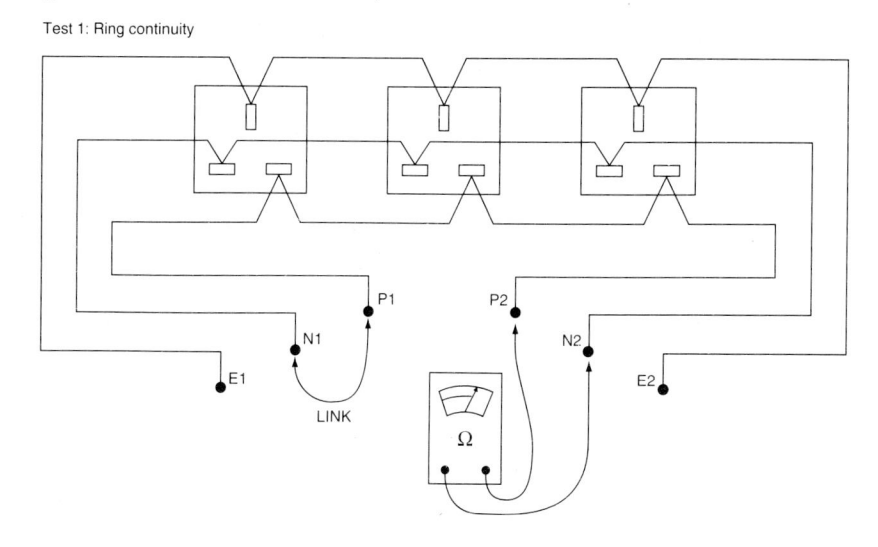

Test 1: Ring continuity

Mark up ends of ring circuit conductors E1, P1, N1, E2, P2, N2. Link P1 to N1 and measure resistance between P2 and N2. Note reading. Repeat test, linking E1 to P1 and note reading between P2 and E2

Test 2: Ring continuity

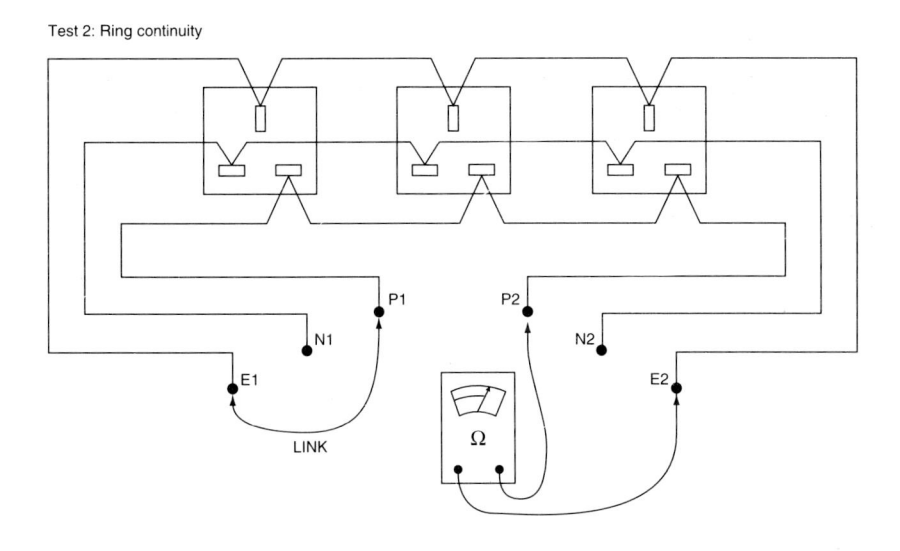

Link together N1 to P2, N2 to P1 and test each socket outlet, noting readings. The readings at each socket outlet should be substantially the same value. Repeat test, linking E1 to P2, E2 to P1 and note readings. This test may also be used for obtaining values of (R1 + R2), see Test 4, when the circuit has a separate cpc.

Test 3: Socket outlets

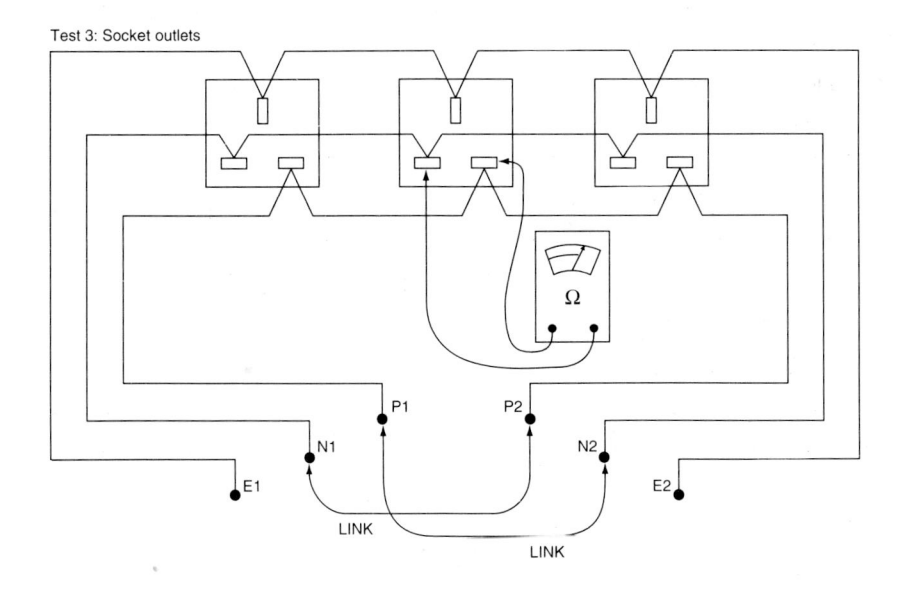

Test 4: Ring continuity R1 + R2 value

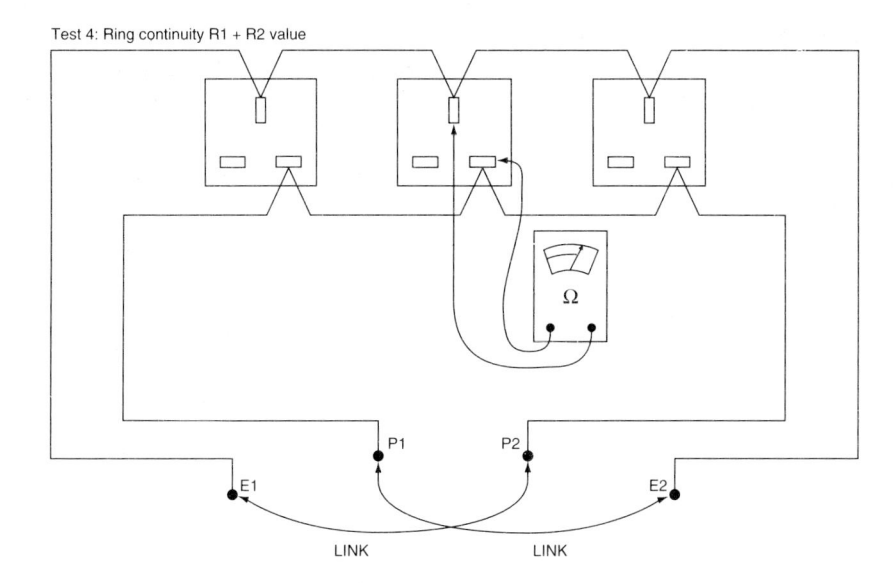

Visual Inspection of Ring Circuit Conductors

An alternative to the above methods for verifying that no interconnection multiple loops have been made in a ring circuit is to inspect each conductor throughout its entire length.

Insulation Resistance

These tests are to verify that the insulation of conductors and electrical accessories and equipment is satisfactory and that the electrical conductors or protective conductor are not short-circuited, or showing a low resistance (which would indicate a deterioration in the insulation of the conductors).

Type of Test Instrument

An insulation resistance tester should be used which is capable of providing a d.c. voltage of not less than twice the nominal voltage of the circuit to be tested (r.m.s. value for an a.c. supply). The test voltage need not exceed the values below when loaded with 1mA.

- 500V d.c. for installations connected to 500V.
- 1000V d.c. for installations connected to supplies in excess of 500V and up to 1000V.

Pre-test Checks

Ensure that neons and capacitors are disconnected from circuits to avoid inaccurate test value being obtained.

Disconnect control equipment or apparatus constructed with semi-conductor devices. These devices will be liable to damage if exposed to the high test voltages used in insulation resistance tests. This requirement will include certain types of rcd.

Ensure lamps and current using equipment are disconnected, and all fuses, switches and MCB's closed.

Typical minimum values of installation resistance are given below:

Circuit Voltage	Up to 500 volts	500 - 1000 volts	Between SELV and LV
Test voltage DC:	500V	1000V	500V
Minimum insulation resistance:	0.5MΩ	1.0MΩ	5.0MΩ

See Table 71A of the IEE Regulations

Insulation Resistance Tests to Earth

All fuses should be in, switches and circuit breakers closed, where practicable any lamps removed, appliances and fixed equipment disconnected. The phase and neutral conductors are connected together at the distribution board and a test is made as illustrated using an insulation resistance tester with test leads being connected between joint phase and neutral conductors and earth.

The reading obtained should not be less than 0.5 Megohms.

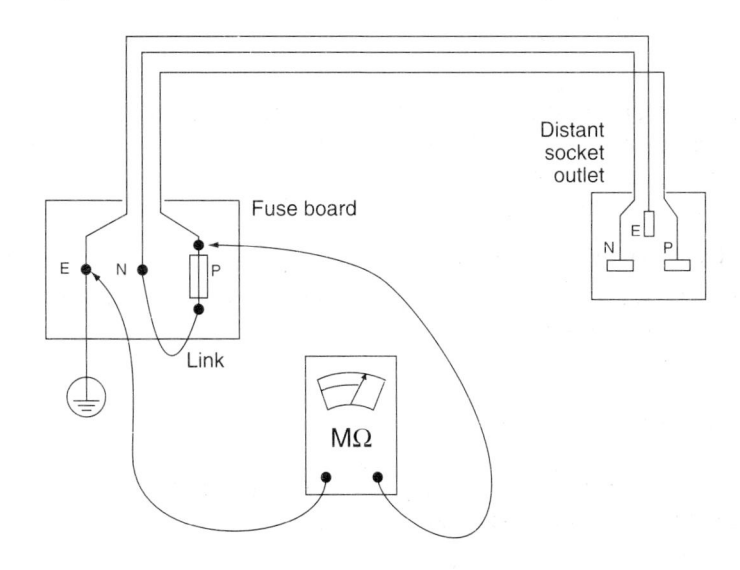

Insulation Resistance Tests Between Poles

All fuses should be in, switches and circuit breakers closed, where practicable any lamps removed, appliances and fixed equipment disconnected. For single-phase circuits the test leads are connected between the phase and neutral conductors in the distribution board.

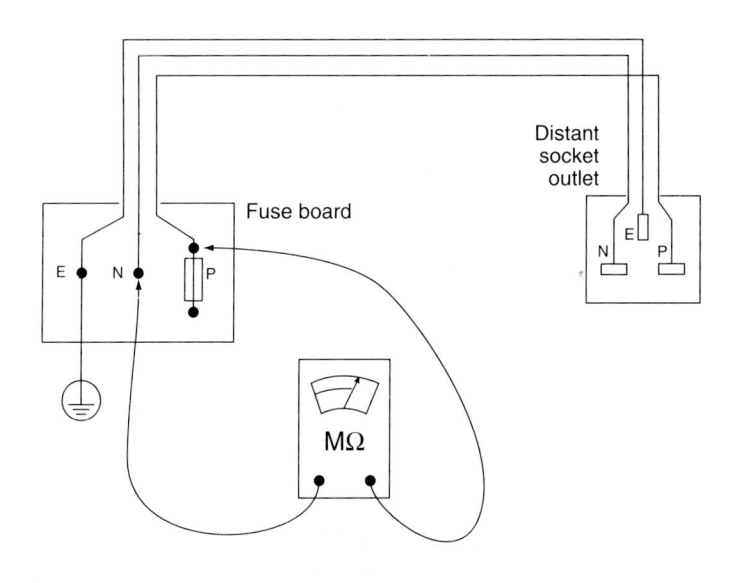

Note: Where any circuits contain two way switching the two way switches will require to be operated and another insulation resistance test carried out, including the strapping wire which was not previously included in the test.

Equipment

When fixed equipment such as boilers have been disconnected to allow insulation resistance tests to be carried out, the equipment itself must be insulation resistance tested between live points and exposed conductive parts.

The test results should comply with the appropriate British Standards. If none, the insulation resistance should not be less than 0.5 megohms.

Warning.
If the fixed equipment contains any electronic components, insulation resistance tests between phase and neutral conductors should not be carried out as this may cause expensive damage to the equipment under test.

Testing Class 2 Equipment

Details physical examination

Insulation test 500V d.c.

Flash test

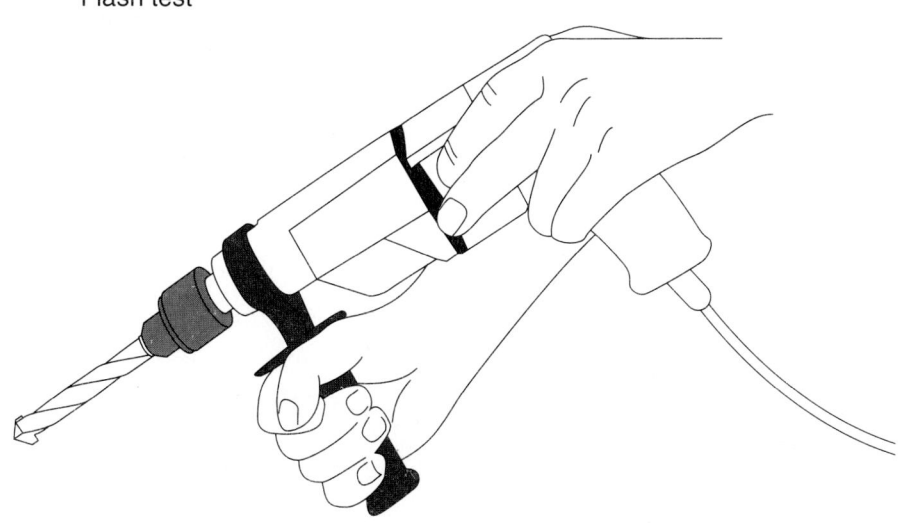

The electrical test performed is the "flash" test. British Standards require a test voltage across basic insulation of 1250V and across the supplementary insulation of 2500V resulting in a combined voltage of 3750V. The same voltage is required if the two separate layers are replaced by a single layer of reinforced insulation. It is generally accepted that a voltage of 3kV is acceptable when repeated periodic testing is performed.

Note: Caution is essential in employing this method. Double insulated equipment frequently incorporates electronic speed control which can be damaged by flash testing. Manufacturer's advice should be sought before applying the test.

The test may only be carried out by a competent person.

Polarity

This test must be carried out to verify that:

(a) All fuses, circuit breakers and single pole control devices such as switches are connected in the phase conductor only.

(b) The centre contact of an Edison type screw lampholder is connected to the phase conductor and the outer metal threaded parts are connected to the neutral or earthed conductor.

(c) Any socket outlets have been correctly installed, i.e. phase pin of 13A socket outlet on right when viewed from the front.

The installation must be tested will all switches in the 'on' position and all lamps and power consuming equipment removed.

A test of polarity can be carried out using a continuity tester as illustrated.

Polarity Test Lighting

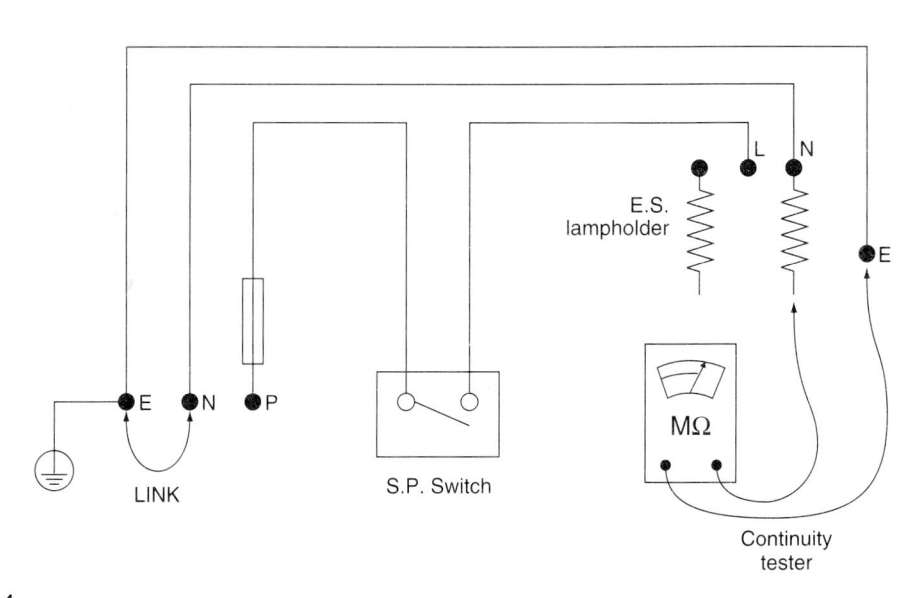

Polarity Test Socket Outlet

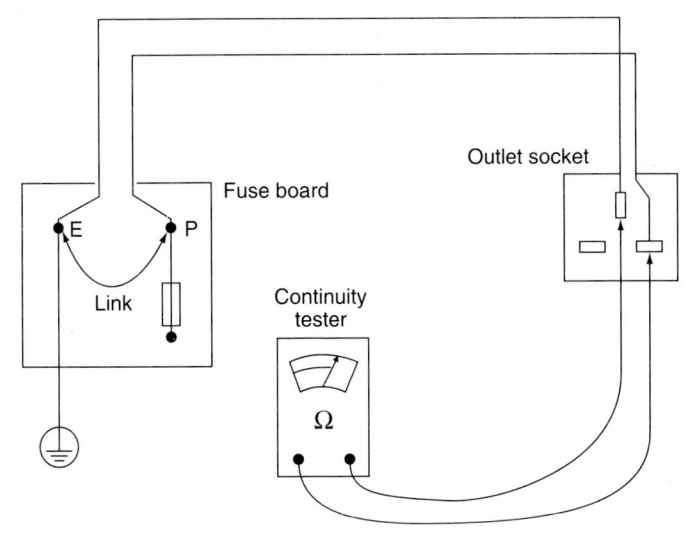

Earth Fault Loop Impedance

The earth fault current loop comprises the following parts, starting at the point of fault for a phase to earth loop.

Circuit protective conductor; the main earthing terminal and earthing conductors; for TN systems the metallic return path (or in the case of TT systems the earth return path); and the path through the earth and neutral point of the transformer, the transformer winding and the phase conductor from the transformer to the point of fault.

The earth fault loop of the TN-S system is illustrated.

Note: The impedance of the earth fault loop is denoted by (Zs).

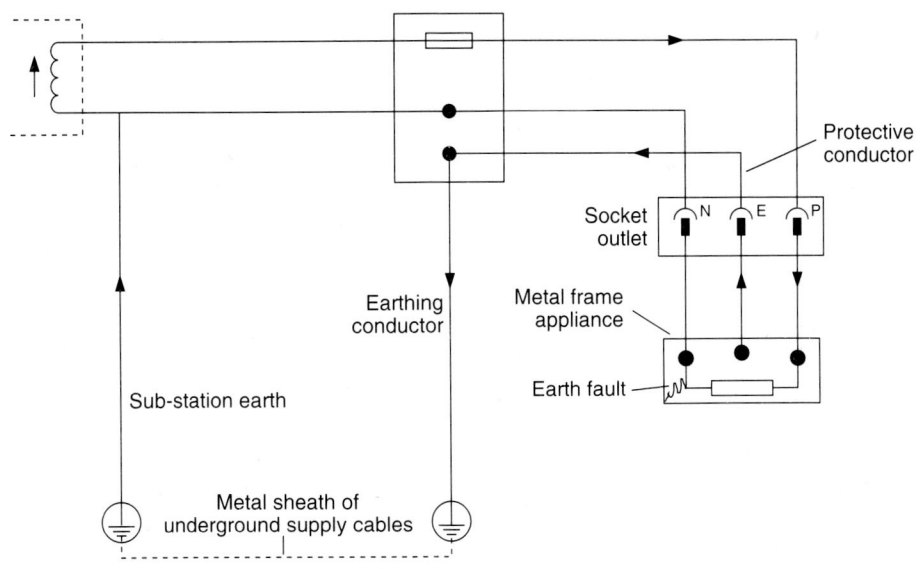

The earth fault loop impedance Zs should always be determined at the most distant point in each final circuit. This is usually carried out at socket outlet circuits and fixed equipment. The values obtained should not exceed those given in Tables 41B1, 41B2 and 41D of the IEE Regulations.

Earth Fault Loop Impedance Testing

The most common method of testing the earth fault loop impedance is by using a phase earth loop impedance tester as illustrated.

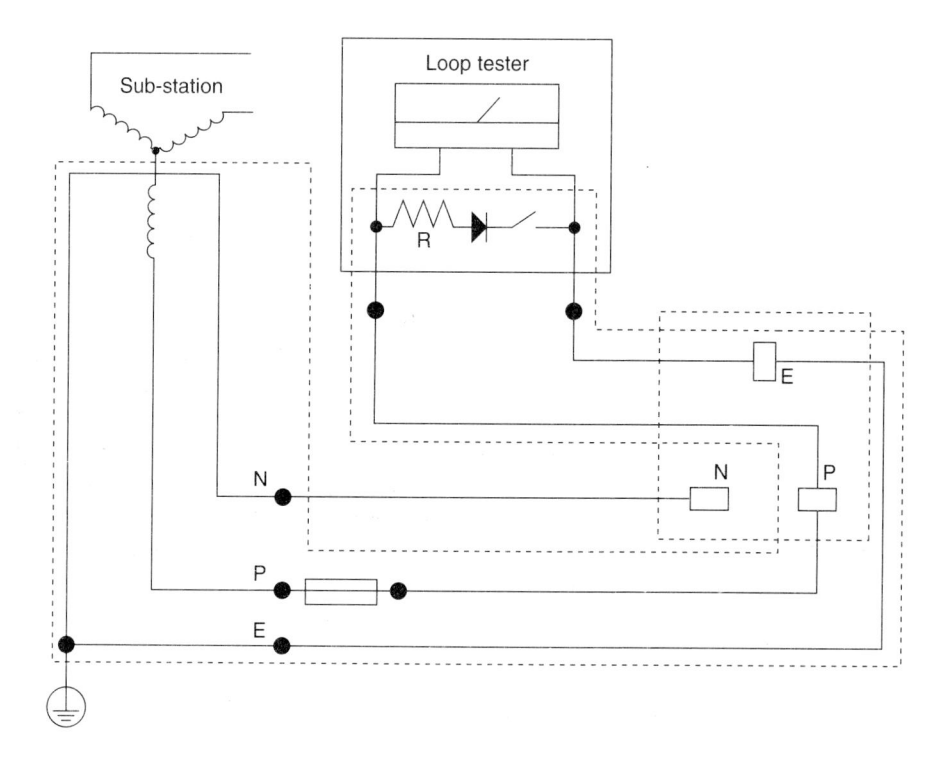

Note: In using such a test instrument, care must be taken to ensure that no ill effects can arise in the event of any defect in the earthing circuit such as would arise if there was a break in the protective conductor of the system under the test. This would prevent the test current from flowing and the whole of the protective conductor system would be connected directly to the phase conductor.

Measurement of the External Earth Fault Loop Impedance Ze

The external earth loop impedance Ze is measured using a phase-earth loop impedance tester as illustrated.

TN-C-S

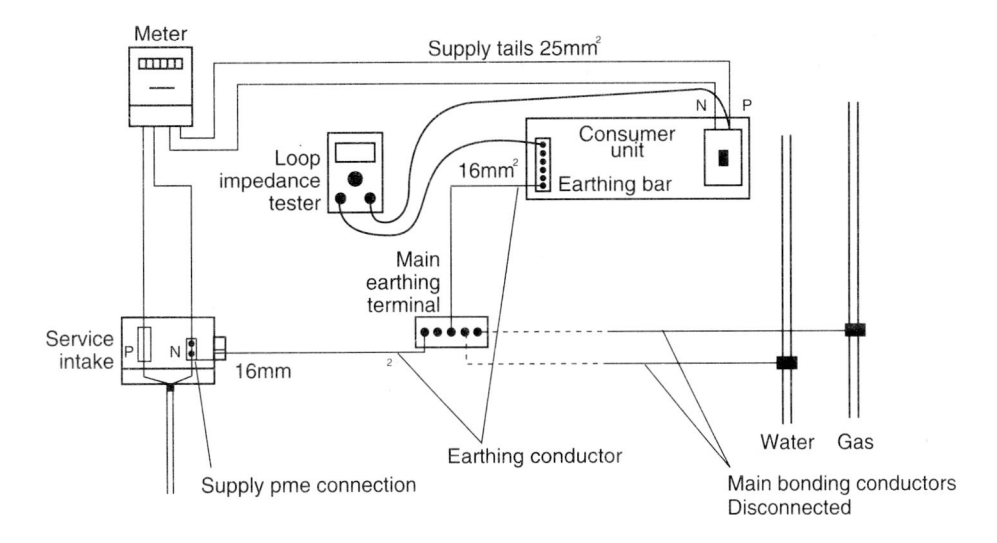

[Diagram: TN-C-S earth fault loop impedance measurement setup showing Meter, Supply tails 25mm², Consumer unit, Loop impedance tester, 16mm² Earthing bar, Main earthing terminal, Service intake, 16mm², Earthing conductor, Supply pme connection, Water, Gas, Main bonding conductors Disconnected]

The value of impedance is measured between the incoming phase supply conductor and the earthing terminal. All main equipotential bonding conductors must be disconnected from the earthing terminal only leaving the earthing conductor connected. The main switch of the consumer unit should be in the open (off) position ensuring all the final circuits are isolated.

Remember after obtaining the value of Z_E, to reconnect the main equipotential bonding conductors.

Operation of Residual Current Devices (BS 4293)

Residual current devices should be tested by simulating appropriate fault conditions, using a test meter.

The test is made on the load side of the circuit breaker, between the phase conductor of the circuit protected and the associated circuit protective conductor, so that a suitable residual current flows. All loads normally supplied through the circuit breaker are disconnected during the test.

Note: When the circuit breaker has a rated tripping current not exceeding 30mA and has been installed to reduce the risk associated with direct contact as indicated in the Note to Regulation 412-06-02, a residual current of 150mA should cause the circuit breaker to open within 40ms.

The effectiveness of the test button or other test facility which is an integral part of the circuit breaker should also be tested by pressing the test button.

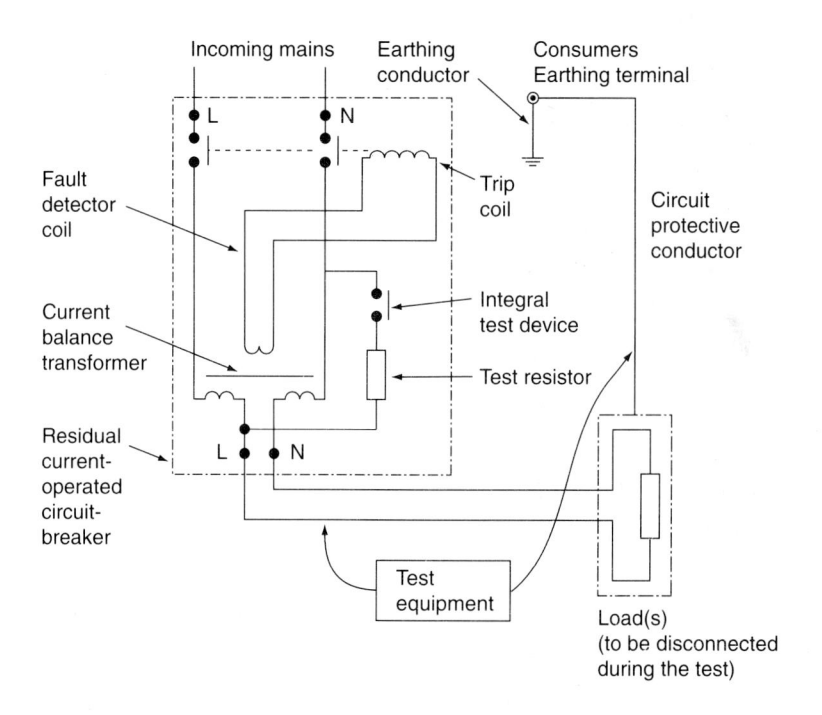

Rcd Testing

Only test meters which give an actual time taken reading should be used. The rcd should be verified by plugging the test meter into a socket outlet which is protected by the rcd. The following fault current settings should achieve the stated result. When selector switch set 0° for phase angle.

Setting	Time to Trip
At 50% of operating current	Does not trip
At 100% of operating current	Device opens within 200ms
At 150mA	Device opens within 40ms

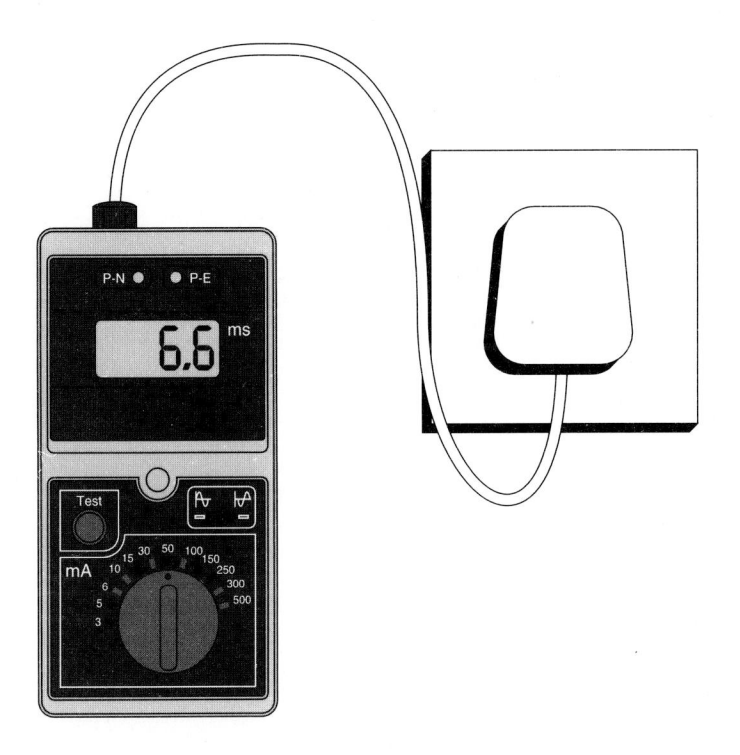

The test should be repeated with phase angle switch set to 180°. The reading should still conform as indicated above.

When making an earth fault loop impedance test on an installation fitted with an rcd this may trip out, especially where the tripping time of the rcd is 30 milliseconds or less. In these circumstances, the rcd should be temporarily short-circuited to permit the earth loop impedance test to be carried out (remembering to remove the short circuit when this test is completed). Since the rcd is inoperative during the time, care should be taken to see that no other tests (or other uses of the circuit) are undertaken until the rcd is back in circuit. Warning notices, etc. should be displayed. Alternatively, a loop impedance tester which is designed not to trip RCD's should be used.

Alterations to Installation

When changes in the use of buildings occur either by change of ownership of use, alterations or additions to the electrical installations often take place. It is important that the person carrying out the electrical work ensures that the work complies with the IEE Regulations and that the existing electrical installation will function correctly and safely.

A completion certificate must be made out and issued for all the work involved in the alteration and any defect found in related parts of the existing installation reported to the person ordering the work by the electrical contractor (or a competent person).

Periodic Inspection and Testing

Periodic inspection and testing of electrical installations and equipment must be carried out to ensure, as far as is reasonably practical:

- safety of persons and livestock against electric shock and burns
- protection of property from damage by fire and heat
- compliance with the requirements of the Regulations

This inspection should consist of careful scrutiny of the installation without dismantling, or by partial dismantling as required, supported by testing.

The periodic inspection and testing of installations is recommended on both the Completion and Inspection Certificate as follows:

	Not Exceeding
Domestic Installations	10 years
Commercial Installations	5 years
Industrial Installations	3 years
Leisure Complexes	1 year
emporary installations on construction sites	3 months

The method of inspection and test should be in accordance with the requirements of the Regulations. An Inspection Certificate must be completed and given to the client.

Note: Shorter intervals between inspections may be necessary depending on the nature of installation. Refer to designers specifications.

Reporting

Following an inspection and test of an installation, a report must be provided by the person carrying out the inspection (or someone authorised to act on his behalf) to the person ordering the work.

Certificates

A combined Completion and Inspection Certificate must be completed and signed by a competent person. It must state that the installation has been:

- designed
- constructed
- inspected and tested

in accordance with the Regulations.

In the case of a small installation, compliance might be achieved by an electrician verifying all these items; a large installation carried out by a major contractor might require the signatures of three persons i.e. the designer, the electrician installing it, and the engineer carrying out the final inspection and test.

FORMS OF COMPLETION AND INSPECTION CERTIFICATE
(as prescribed in the IEE Regulations for Electrical Installations)

(1) (see Notes overleaf)

DETAILS OF THE INSTALLATION

Client:

Address:

DESIGN

I/We being the person(s) responsible (as indicated by my/our signatures below) for the Design of the elretical installation, particulars of which are described on Page 3 of this form CERTIFY that the said work for which I/we have been responsible is to the best of my/our knowledge and belief in accordance with the Regulations for Electrical Installations published by the Institution of Electrical Engineers, 16th Edition, amended to (3.) (date . . . .) except for the departures, if any, stated in this Certificate.

The extent of liability of the signatory is limited to the work described above as the subject of this Certificate.

For the DESIGN of the installation:

Name (In Block Letters):   Position:

For and on behalf of:

Address:

(2.) Signature: (3.) Date

CONSTRUCTION

I/We being the person(s) responsible (as indicated by my/our signatures below) for the Construction of the electrical installation, particulars of which are described on Page 3 of this form CERTIFY that the said work for which I/we have been responsible is to the best of my/our knowledge and belief in accordance with the Regulations for Electrical Installations published by the Institution of Electrical Engineers. 16th Edition, amended to (3.) (date . . . .) except for the departures, if any, stated in this Certificate.

The extent of liability of the signatory is limited to the work described above as the subject of this Certificate.

For the CONSTRUCTION of the installation:

Name (In Block Letters):   Position:

For and on behalf of:

Address:

(2.) Signature: (3.) Date:

INSPECTION AND TEST

I/We being the person(s) responsible (as indicated by my/our signatures below) for the Inspection and Test of the electrical installation, particulars of which are described on Page 3 of this form CERTIFY that the said work for which I/we have been responsible is to the best of my/our knowledge and belief in accordance with the Regulations for Electrical Installations published by the Institution of Electrical Engineers, 16th Edition, amended on (3.) (date . . . .) except for xdepartures, if any, stated in this Certificate.

The extent of liability of the signatory is limited to work described above as the subject of this Certificate

For the INSPECTION AND TEST of the installation:

Name (In Block Letters):   Position:

For and on behalf of:

Address:

I RECOMMEND that this installation be further inspected and tested after an interval of not more than . . . . . years.(5.)

(2.) Signature: (3.) Date:

(6) page 1 of . . . . . . pages

Reproduced from the IEE On-Site Guide with acknowledgements to the Institution of Electrical Engineers

PARTICULARS OF THE INSTALLATION
(Delete or complete items as appropriate)

Type of Installation   New/alterations/addition/to existing installation

Type of Earthing (312-03):   TN-C   TN-S   TN-C-S   TT   IT
(Indicate in the box)   ☐   ☐   ☐   ☐   ☐

Earth Electrode:
     Resistance . . . . ohms
     Method of Measurement . . . .
     Type (542-02-01) and Location . . . .

Characteristics of the supply at the origin of the installation (313-01):

     Nominal voltage . . . . volts

     Frequency . . . . Hz

     Number of phases . . . .

ascertained	determined	measured

     Prospective short-circuit current . . . . .kA

     Earth fault loop impedance (Z_E) . . . . ohms

     Maximum demand . . . . A per phase

     Overcurrent protective device -Type BS . . . . Rating . . . . A

Main switch or circuit breaker (460-01-02): Type BS . . . . Rating . . . . A No of poles . . . . . .

     If an r.c.d., rated residual operating current I∆ . . . . .mA.)

Method of protection against indirect contact:

                                                                                          ☐

1. Earthed equipotential bonding and automatic disconnection of supply

or   ☐

2. Other   (Describe) . . . .

Main equipotential bonding coductors(413-02-01/02, 546-02-01): Size . . . . mm₂

Schedule of Test Results: Continuance . . . . pages

Details of departures (if any) from the Wiring Regulations (120-04,120-05) . . . .

Comments on existing installation, where applicable (743-01-01): . . . .

(6) page 3 of . . . . . . pages

Reproduced from the IEE On-Site Guide with acknowledgements to the Institution of Electrical Engineers

INSTALLATION SCHEDULE

CLIENT: .... JOB No: ....

....

....

ADDRESS: ....

....

DESCRIPTION OF WORK PERFORMED:

SUPPLY DETAILS		EARTHING SYSTEM					EXTERNAL IMPEDANCE Ze			PSC	kA	
Circuit	No. of points	Fuse / MCB	Rating amps	Cable size mm²	length m	CPC mm²	Zs Ω	Ins Res MΩ	Polarity	RCD mS	Ring Cont	Remarks
1.												
2.												
3.												
4.												
5.												
6.												
7.												
8.												

TEST INSTRUMENT DETAILS	MAKE	MODEL	SERIAL No
Continuity			
Insulation Resistance			
Polarity			
Earth Loop Impedance			
R.C.D.			
Other (State)			

COMMENTS ON EXISTING INSTALLATION

NAME (BLOCK LETTERS) .... POSITION: ....

FOR AND ON BEHALF OF: .... SIGNATURE: ....

ADDRESS: .... DATE: ....

....